Terra Nostra

Jeffrey S. Murray

Terra Nostra

THE STORIES BEHIND CANADA'S MAPS

1550–1950

From the collection of
Library and Archives Canada

McGILL-QUEEN'S UNIVERSITY PRESS  SEPTENTRION

Published by les éditions Septentrion in co-operation with Library and Archives Canada, and Public Works and Government Services Canada.

Les éditions du Septentrion thanks the Canada Council for the Arts for its support. We are also grateful for the financial support of the Government of Canada through its Book Publishing Industry Development Program (BPIDP).

FRONT COVER:
Theodor de Bry, *America sive novvs orbis respectv Evropaeorvm inferior globi terrestris pars*, 1596 (page 184)

BACK COVER:
Unknown artist, *Niagara Falls Park and River Railway*, 1900 (pages 78–79)

ENDPAPERS:
Paolo Forlani, *Il Mondo*, Venetiis, Ioan. Francisci Camotii aeries formis, 1560, NMC 97953

PROJECT DIRECTORS:
Claire Rochon (Library and Archives Canada) and Denis Vaugeois (Septentrion)

EDITING AND PROOFREADING:
Marcia Rodríguez and Michèle Brenckmann

DESIGN AND LAYOUT:
Folio Infographie

Library and Archives Canada Cataloguing in Publication

Murray, Jeffrey S.

    Terra nostra : the stories behind Canada's maps, 1550-1950

    Co-published by: Library and Archives Canada, Public Works and Government Services Canada.
    Includes bibliographical references and index.

    ISBN 2-89448-453-4 (Éditions du Septentrion)
    ISBN 0-660-19496-1 (Library and Archives Canada)

1. Canada – Historical geography – Maps. 2. Cartography – Canada – History. 3. Library and Archives Canada – Map collections – History.
I. Library and Archives Canada. II. Canada. Public Works and Government Services Canada. III. Title.

GA471.M87 2006   912.71   C2006-940235-3

Legal Deposit – 1st quarter 2006
Bibliothèque nationale du Québec
Library and Archives Canada
ISBN 2-89448-453-4

Les éditions du Septentrion
1300, avenue Maguire
Sillery, Quebec
G1T 1Z3

DISTRIBUTION IN CANADA:
McGill-Queen's University Press
c/o Georgetown Terminal Warehouses
34 Armstrong Avenue
Georgetown, Ontario
L7G 4R9

PRINTED IN CANADA BY FRIESENS

Contents

Acknowledgements

Some chapters in this book have appeared previously—sometimes in substantially different form and under a different title—in the following magazines: "Passage to the Orient" in *This Country Canada* and *Mercator's World*; "Modelling Québec" in *Mercator's World*; "Going for Klondike Gold" in *Mercator's World*; "Deville's Topographic Camera" in *Legion Magazine*; and "Weapons of the Great War" in *Legion*. I would like to extend my sincere thanks to the editors of these magazines for their support in this publication.

A work of this magnitude would not have been possible without the persistent dedication of a number of colleagues at Library and Archives Canada. I am particularly indebted to Ian Wilson, who initially suggested the project; to Gabrielle Blais and Jean-Stephen Piché, who diligently negotiated the bureaucratic minefield behind the scenes to make it possible; and to my colleagues in the Web Content and Services Division, who waited patiently for my return while my duties took me elsewhere.

I would like to thank Eric Boudreau, Kim Dubois, Dave Knox, Chris Lachance, and Sophie Dazé, who prepared the digital images for publication; Suzanne Hotte-Guibord and Kevin Joynt, who were instrumental in finding specialized references; Bruce Weedmark and Pat McIntyre, who always said "yes" to my many requests; Carole Cloutier, who patiently researched and obtained copyright approvals; Terry Cook, the late Bruce Wilson, Danny Moore, and Doug Whyte, who over the years provided a magnificent environment for contemplating the past; and, finally, Louis Cardinal and Betty Kidd, who willingly took a novice archivist under their wing and introduced him to the wonderful world of early cartography.

I also wish to thank the individuals who participated in the editing and production of this book: Claire Rochon, whose guidance and support enabled the publication to proceed smoothly through all its stages; Marcia Rodríguez and Michèle Brenckmann for the English text, and Myriam Afriat for the French text—whose editing skill and insight greatly enhanced the manuscript; Julie Desgagné, whose excellent translation brought out the full meaning and tone of the text; Josée Lalancette, whose creative layout and design make this volume so pleasing to the eye; and Denis Vaugeois, whose vision and perspective helped to shape the final version of this book.

My greatest debt, however, is to my wife, Dana-Mae. She has always been my steady navigator and companion through life.

For my parents

Gerard Mercator's famous map of the North Pole was published posthumously by his son in 1595. Canada is the land mass along the left border; the North Pole is the massive rock in the centre. The Flemish cartographer surrounded the pole with four large islands. According to the theories of the day, the waters separating the islands were thought to be drawn northward to the pole by the earth's rotation. It was thought that this action created a current strong enough to keep the four straits free of ice all year.

Gerardus Mercator, *Septentrionalium terrarum descriptio,* Rupella, Typographus sumptibus haeredum Gerardi Mercatoris Rupelmundi, 1595. From *Atlas sive cosmographicae meditationes de fabrica mundi et fabricata figura,* NMC 16097

Foreword

CANADA'S MAPS ARE PRECIOUS. They are a vivid record of accomplishment in the vast territory of Canada; they document the labours and accumulated knowledge of countless individuals over four centuries. Beautiful and detailed, and inevitably inspiring, they reveal much about the development of this land and of the diversity of its people. Maps assemble and give coherence to what was learned from our First Nations; they record the experiences and findings of those who travelled, explored, and later surveyed our land. They provide the context in which our social, economic, and political development unfolded, and they remain essential to understanding the Canadian experience. Many are delightful to the eye; all warrant careful study and offer insight to the inquisitive mind.

The cartographic holdings of Library and Archives Canada, gathered through more than a century of active collecting, include more than 1.7 million maps, plans, and charts. The collection spans the full gamut of mapmaking—from extensive documentation on early cartography and hydrography to current electronic mapping. This is a national resource, carefully preserved for the future, but increasingly known and accessible today. More than 4,000 maps are already available on our website, together with digital collections and virtual exhibitions that highlight all aspects of our collection. Available to educators, students, researchers, and all who are interested, this electronic presence will continue to grow over time.

Library and Archives Canada is committed to making its extraordinary multi-media collection relating to Canada available to a broad audience. This illustrated volume on the development of cartography in Canada is a fine example of one such initiative. I extend a special thanks to its author, Senior Archivist Jeffrey Murray, whose passion for cartography, expertise on Canada's mapping history, and extensive knowledge of our collection have culminated in this beautiful narrative on Canada's maps.

As you delve into *Terra Nostra*, be prepared to be enthralled! Discover some of our most valued cartographic pieces; share in their "making" and raison d'être; appreciate their detail and elegant line. Gain a greater understanding of the role that maps have played in helping us to conceptualize our world and our place in it.

Meet the cartographers and surveyors; the explorers and adventurers; the risk takers and creators: the people who have contributed to Canada's maps and have played an important role in our country's history. Their stories are extraordinary. Their works are magnificent.

Terra Nostra is being released as we celebrate with Natural Resources Canada the one-hundredth anniversary of our country's first national atlas. Many of the accomplishments and technological innovations that are presented in this book have issued from this federal government department, and I am pleased that we can highlight them here as part of our country's extraordinary documentary heritage.

I am proud to showcase our national cartographic treasures and to share their stories with you.

IAN E. WILSON
Librarian and Archivist of Canada

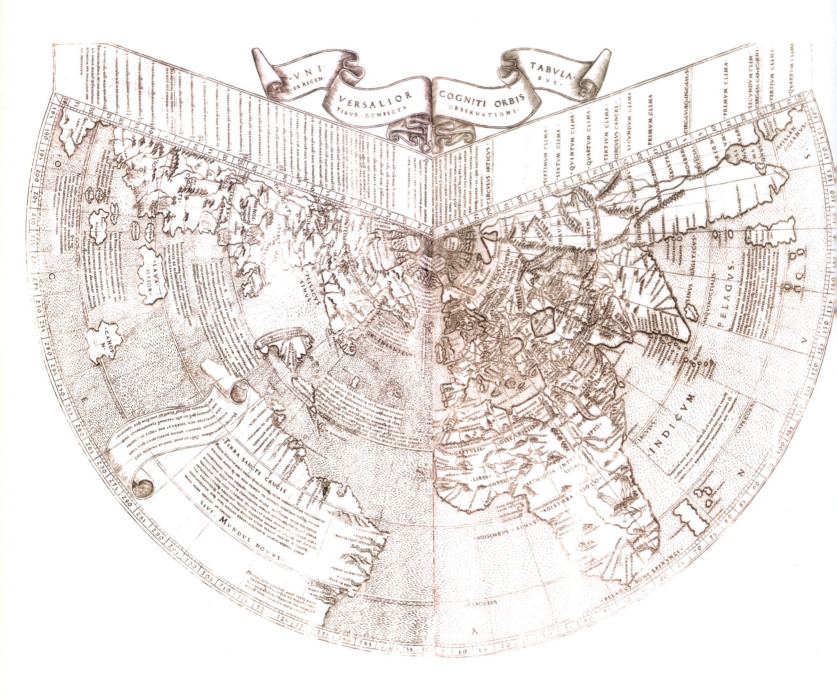

Johannes Ruysch's map of the world, published in a 1508 edition of Ptolemy's *Geographia*, is the earliest original document relating to Canada in the Library and Archives Canada collection. The eastern limit of Canada, the islands of Newfoundland and Cape Breton, appear here with their Latin name *Terra Nova*, and are shown as a peninsula of Asia, due west and on the same parallel as France.

Johannes Ruysch, *Universalior cogniti orbis tabula ex recentibus confecta observationibus*, Rome, Impressum per Bernardum Venetum de Vitalibus, expensis Evangelista Tosino Brixiano Bibliopola, 1508. From *In hoc opere haec continentvr geographiae Cl. Ptolemaei a plurimis uiris utriusque linguae doctiss.*, NMC 19268

Preface

I LIKE TO THINK THAT I HAVE the best job in the world. I spend my working days with old maps in one of our country's greatest national repositories—Library and Archives Canada. If I am not in one of our humidity and temperature-controlled vaults, looking for maps to scan for our website, I might be at my desk, building a specialized guide to the 1.7 million maps in our holdings. I might also be found in the reference room, assisting a researcher on a project; in a government office, preparing a transfer of maps that are no longer of service to the department; or perusing a dealer's sale catalogue for old maps to add to the national collection. My colleagues and I are stewards of a valuable national asset—one that our generation has inherited from our predecessors—one that we will add to and pass on to future generations.

At Library and Archives Canada we are committed to making our unique and engaging treasures available to as broad a public as possible, subject to technological limitations and the imperatives of preservation. In many ways, this book is an outgrowth of that commitment. It covers four centuries of Canadian mapmaking, from approximately 1550 to 1950. It is a history of some of the more important maps that have allowed us to come to grips with the geographical diversity of *terra nostra*—our land—as we forged a nation from sea, to sea, to sea.

The maps considered in this volume were produced before the development of satellite imaging and the dominance of computerized, geographical information systems in the mapmaking professions. It was a period marked by profound technological change and an insatiable quest for greater and greater knowledge of distant lands—a quest that resulted in the creation of new sciences and a complete re-examination of our understanding of the universe and humankind's role in it. Who were the people who made the maps that we revere in our collection today? How were they made? Why were they made? How did this form of communication affect the lives of Canadians? These are some of the questions addressed in this book.

Parts of this story have already been told, but not from the historical perspective offered by this volume. Until now, much of the narrative of Canada's cartographic history has fallen on the shoulders of former practitioners in the mapmaking and surveying professions. Their accounts are usually personal, and though their writings add a human dimension to our cartographic legacy, they are difficult to find and often contain specialized terminology. To some degree, Canada's antiquarians have also shown a keen interest in early cartography and many of their compilations are quite beautiful; however, antiquarians treat maps as objects of veneration rather than artifacts with a fascinating history in their own right. In addition, their discussions generally end around the mid nineteenth century; the period during which some of the most remarkable developments in cartography were introduced—the last 150 years of mapmaking—is not covered.

Canada's historians have written fascinating accounts of the early exploration of our country and have provided wonderful chronicles of our attempts at nation building. But the maps that were a product of these great endeavours are either not represented or treated as little more than artwork to illustrate broader socio-economic themes. Regrettably, historians have not concentrated on the problems faced by cartographers as they attempted to define our vast *terra incognita*, nor on the role that maps have played in the development of our country.

With 2006 marking the one-hundredth anniversary of Canada's first national atlas, it is a perfect occasion to recount some of the stories behind Canada's maps. Anniversaries are customarily a time for reflection and a time to evaluate past accomplishments. The 1906 *Atlas of Canada*—the second national atlas in the world after Finland and one that has been copied many times by many countries—represents a significant development in Canada's ability to deal with its vast landscape and promote national aspirations. By focusing on Canadian advances in communication, transportation, and natural resource development, and on the country's ethnic diversity, the 1906 atlas was a celebration of many of the values that we still hold dear today.

As a cartographic archivist and a chronicler of the past, I would like to see this anniversary and its celebration become part of Canada's social fabric. It is only with an understanding of our shared history, in all its complexity and diversity, that we can begin to address the serious challenges of our future. The stories behind Canada's maps are very human; they are full of rivalries, hard work, brilliant insight, and technological innovation. Most of all, they reveal a passion for the exploration and discovery of our corner of the world. ∾

JEFFREY S. MURRAY
Senior Archivist
Library and Archives Canada

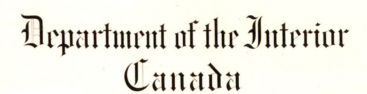

Published one year after the entry of Alberta and Saskatchewan into Confederation, the 1906 *Atlas of Canada* was a celebration of Canadian achievement. This title page shows the crests of all the provinces. Canada had finally achieved its dream as a nation *a mari usque ad mare*—from sea to sea.

Canada, Dept. of the Interior, *Atlas of Canada*, Ottawa, Dept. of the Interior, 1906, cover page.

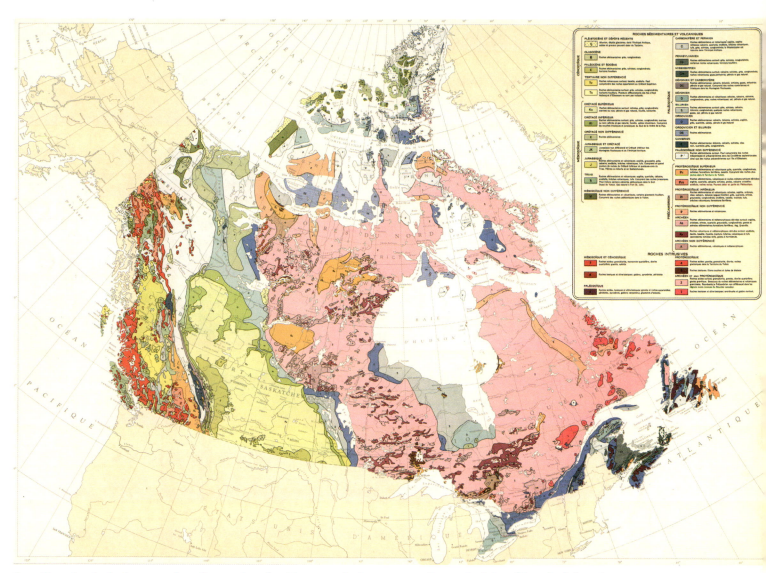

Canada's national atlas has gone through a number of revisions over the last 100 years. A second edition of the atlas followed quickly in 1915, which was a year of exceptionally high immigration to western Canada. This plate showing Canada's bedrock geology was compiled in 1956 by the Geological Survey of Canada for the third atlas, which was published in English in 1957 and for the first time in French in 1958. The fourth atlas, retitled *The National Atlas of Canada*, was issued in 1974. The 93 maps in the fifth edition were produced in serial format starting in 1978. This marked the first time that remotely-sensed data from satellites was used in the atlas. *The National Atlas on the Internet* was released in 1994; it was one of the first online atlases in the world. The sixth edition of the Atlas was launched on the Internet in 1999.

Canada, Ministère des mines et des relevés techniques, *Atlas du Canada*, Ottawa, Ministère des mines et des relevés techniques, Commission géologique du Canada, 1958

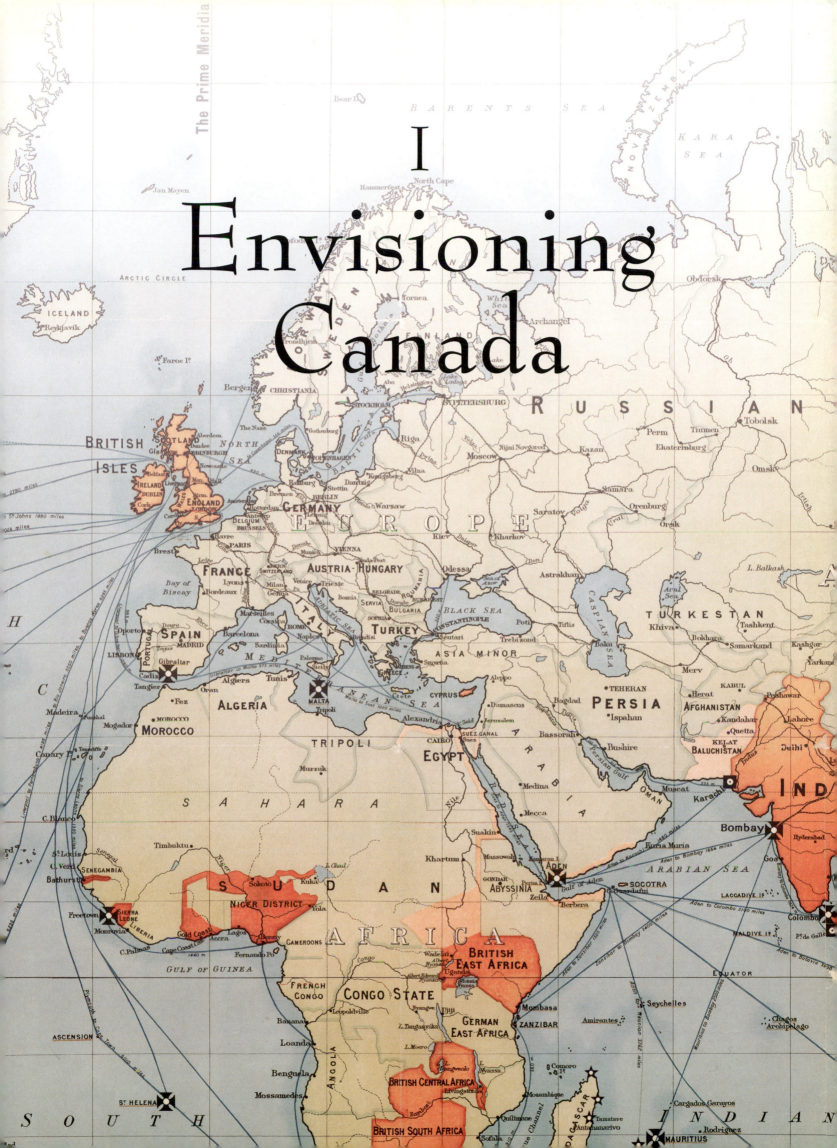

I
Envisioning
Canada

MAPS ARE POWERFUL TOOLS that reflect our view of the world. Part of this power comes from our natural willingness to accept them as they are. Where we are taught from childhood to question written text—to read between the lines—we tend to be less critical of the information conveyed through maps. "Add a map," explains geographer Mark Monmonier, "and the idea not only sounds good but it looks good too." In other words, maps are valuable tools of persuasion that enable their creators to present a singular vision of the world. When users share this vision, the inherent subjectivity in maps will reinforce their understanding of the world, but when the vision is different, the resulting image can be a cause for concern, even outrage. Clearly, maps work at several levels simultaneously and contain different meanings for different people.

With mass media now playing such a vital role in the control of society, any group that legitimates its ideas through maps and other forms of communication not only becomes better known and understood but ultimately may also succeed in obtaining better

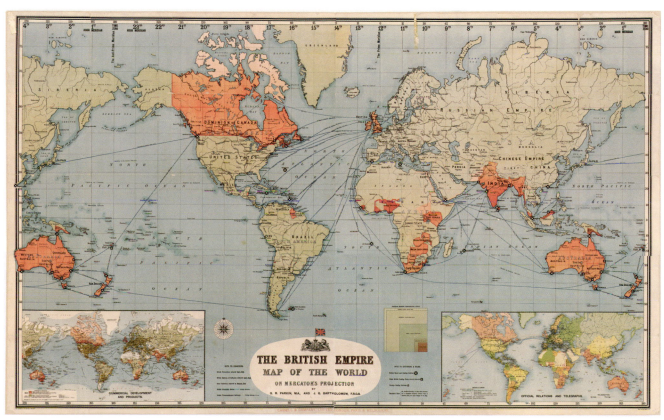

A New Brunswick school teacher, George R. Parkin, produced this wall map of the British Empire for schools and libraries in 1893. For Parkin, advances in shipping and telegraphic communication had transformed the Empire into a single geographical entity with Canada as the keystone holding the Pacific and Atlantic nations together. His map envisioned this new role for Canada by placing it top centre between two Australias.

George Robert Parkin, *The British Empire map of the world on Mercator's projection*, Edinburgh, J.G. Bartholomew, 1893, NMC 16992

The Department of Indian Affairs published this map in 1891 to show the system of Indian reserves it had set up across the country, not to give First Nations a homeland, but to control the Aboriginal population and pave the way for commercial agriculture. With First Nations placed on reserves, such maps demonstrated to the world that Canada was a safe place to invest and settle.

Canada, Dept. of Indian Affairs, *Map of the Dominion of Canada showing Indian reserves ...*, Ottawa, Dept. of Indian Affairs, 1891, NMC 117506

TREATY 3. Continued.

No	Name
34	Sabaskong Bay.
	a. White Fish Bay.
	b. Shoal Lake.
	c. N.W. Angle River.
	a. Naongashing.
	b. Osabukong.
	c. Sabaskong.
	d. East of 35c.
35	e. Little Grassy R.
	e2. S.E. from 35c.
	f. Sabaskong Bay.
	g. Big Grassy River.
	h. Sabaskong Bay.
	j. Near 35B, 2 Islands.
36	Buffalo Point.
37	Big Island.
37	Rainy River.
37a	Shoal Lake.
38	Oak Island & N.W.A.
38a	North West Angle.
38	Near Rat Portage.
38a	Clear Water Bay. 5 Isls.
39	Shoal Lake.
39	Shoal Lake.
40	Islands in Shoal Lake.

INDEX TREATY 3.

No	Name
10	Rainy River.
11	Do. do.
12	Do. do.
13	
14	Hungry Hall.
15	
16	Rainy Lake.
17	Rainy Lake.
18	Rainy Lake.
19	
21	English River.
22	Wabaskang.
	31. Hell Lacs.
	a. Lone Fiested River.
23	2. Mouth of Seine R.
	32. River Seine.
24	Kaskagumot R.
25	Pequaquam Lake.
26	6. Rainy Lake.
27	Wabigoon & Eagle Li.
28	Lac Seul.
29	Winnipeg River.
29	Swan Lake.
30	One Man's Lake.
30	Sabaskasing.
	1. Lake of the Woods.
	2. North West Angle.
	c. Lake of the Woods.
31	a. Big Island Lac W.
	b. Big Island.
	f.
	g.
	h.
31	Shoal Lake.
	a. White Fish Bay Isls.
32	Yellow Girl Bay.
	c. Sabaskong Bay.
33	a. White Fish Bay.
	b. North West Angle.

INDEX QUEBEC.

No	Name
1	Mann.
2	Maria.
3	Belsiamites.
4	Viger.
5	Caiatchouan.
6	Ouiacuna.
7	Lorette.
8	Quarante Arpent.
9	Becmont.
10	St Maurice R. Unsurveyed
11	Becancour.
12	River S. Francis (2)
13	Coteraine.
14	Caughnawaga.
15	St Regis.
16	Oka.
17	Doncaster.
18	Maniwaki.
19	Temiscaming.
20	Islands in St Lawrence.

INDEX ONTARIO.

No	Name
1	Maganetawan.
2	Henry Inlet.
3	Point Grondine.
4	White Fish River.
5	Spanish River.
6	Serpent River.
7	Mississaga.
8	Doris.
9	
10	Nipissing.
11	Wahauanitaping.
12	Thessalon.
13	French River.
14	Garden River.
15	Surrendered.
16	Parry Island.
17	Shawanaga & 17a
18	Tamaganing.
19	Cockburn Island.
20	Shesheguaning.
21	Dridgewong.

INDEX ONTARIO. Continued.

No	Name
22	West Bay.
23	Sucker Creek.
24	Shegaiandah.
25	Sucker Lake.
26	Manitoulin Id.
27	Cape Croker.
28	Chiefs Point.
29	Saugeen.
30	Christian Ids.
31	Gibson.
32	Rama.
33	Georgina Id.
34	Scugog.
35	Mud Lake.
36	Rice Lake.
37	Alnwick.
38	Tyendinaga.
39	Golden Lake.
40	Tuscarora.
41	Delaware.
42	Caradoc.
43	Aux Sable.
44	Kettle Point.
45	Sarnia.
46	Walpole Id.
47	Oxford.
48	Michipicoten.
49	Grey Cap.
50	Pic River.
51	Pays Plat.
52	Port William.
53	Red Rock.
54	McIntyre Bay.
55	Gull River.
56	Island Point.
57	Jackfish Island.
58	Long Lake.

INDEX NEW BRUNSWICK.

No	Name
1	Indian Point.
2	Eel Ground.
3	Eel River.
4	Miramichi Res.
5	Part of No 10.
6	French Village.
7	Sec A & 4.
8	Big Hole Tract.
9	Tabusintac.
10	St Basil.
11	Nquisguit.
12	Cancus.
13	Pokemouche.
14	Burn. Church.
15	Richibucto.
16	Buctouche.
17	Botsford.
18	The Brothers.
19	Cameros River.
20	Tobique.
21	Great Bend.
22	St Croix.
23	Woodstock.
24	St Marys.

INDEX NOVA SCOTIA.

No	Name
1	Middle River.
2	Whykokomagh.
3	Malagawatch.
4	Eseasoni.
5	Chapel Island.
6	Bear River.
7	Cayumcoga Lake.
8	Near Liverpool Road.
9	Cayumcoga Lake.
10	Penhook.
11	Medway.
12	
13	Shubenacadie Lake.
14	Indian Brook.
15	Sambro.
16	Ingrams River.
17	Middle River.
18	Ship Harbour.
19	Pennalts.
20	New Ross.
21	Gold River.
22	Cumberland.
23	Little River 3 Res.
24	Picton Harbour.
25	Marquite River.
26	Port Rowl. Not defined
27	Mill Brook.
28	Sydney.
29	Carribee Marsh.
30	Cow Bay.
31	Merigomish.
32	Cornwallis.

PRINCE EDWARD ISLAND

No	Name
1	Lenox Island.
2	Morel.

Department of Indian Affairs.

MAP

of the

DOMINION OF CANADA

SHOWING INDIAN RESERVES

Scale of Statute Miles

Projection on Oblique Secant Cylinder

To accompany Annual Report of

1891

Deputy Supt Genl. of Indian Affairs.

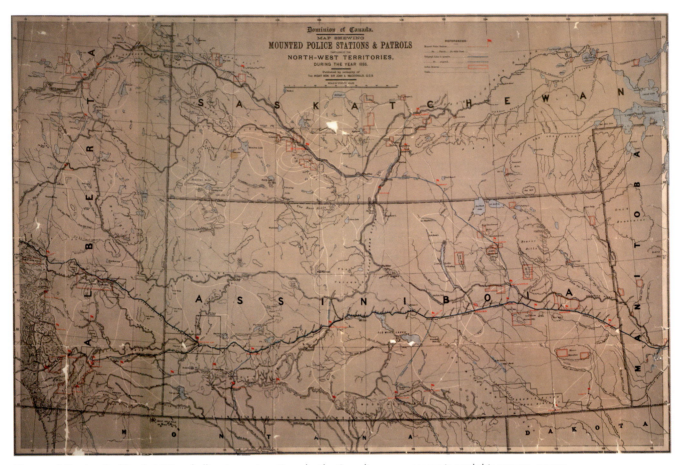

The year following the bloody Métis rebellion in western Canada, the Canadian government issued this map to assure a nervous public that the situation was well under control. It shows the distribution of federal police stations and patrol routes across western Canada. Following its introduction in 1886, the map was updated and re-issued on a regular basis to 1906, when the newly created provinces of Alberta and Saskatchewan assumed their own policing responsibilities.

Canada. Dept. of the Interior, *Map shewing mounted police stations & patrols throughout the North-West Territories*, Ottawa, Dept. of the Interior, 1886, NMC 11793

access to the distribution of power, prestige, and wealth. It is not surprising then, that the ideas and thoughts society's competing forces transmit through maps are neither random nor static, but embedded deeply within human social purpose.

This section examines three instances in which maps have been used as instruments of hope, control, and persuasion. The first chapter looks at Europe's search for the Northwest Passage, that magical highway to the legendary riches of Cathay and the Far East. It was a search that turned the New World into little more than a brief interlude in Europe's westward expansion to Asia and transformed every promising waterway into a powerful magnet attracting exploration. Europe's tenacity and faith in the existence of an all-water route that would lead around or through the continent coloured much of its under-

standing of North American geography for over three hundred years. The Northwest Passage was a cartography of hope.

For centuries, Europe's monarchs asserted sovereignty over North America by simply shifting imaginary borders or renaming the geographical features on their maps: far easier to move lines on a map than armies across battlefields halfway around the world. No wonder that one of the first acts of Britain's occupational army following the fall of New France in 1759 was the construction of a large-scale map of the St. Lawrence corridor. As described in the second chapter, the Murray map had a practical application if armed conflict should ever occur again. However, it also had its symbolic aspects. As teams of red-coated military personnel searched family farms and conducted a census of able-bodied men and household firearms, the habitant farmers surely would have viewed the project as the act of a conquering army on a subjected people. The Murray map was a cartography of control.

The third chapter looks at a series of atlases published by the federal government in the closing years of the nineteenth century. These atlases were introduced in an attempt to change British attitudes to the western Prairies and were part of a much larger advertising campaign that saw immigrants from Europe open homestead lands across the west. And it worked! Perceptions changed, immigrants arrived, and the naked landscape was turned into one of Canada's most productive agricultural regions. These publications helped Canada to create a new identity in the international arena. The atlases were a cartography of persuasion. ❧

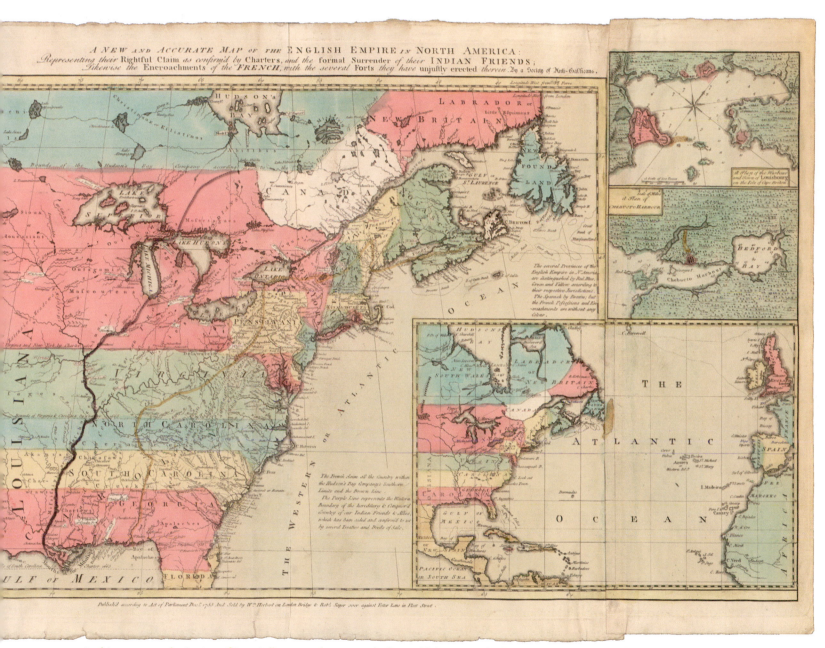

In this 1755 map, the Society of Anti-Gallicans tried to persuade the world that Britain had sovereignty over much of North America. Their claim projected into the lands of the Iroquois confederacy and reduced New France (shown without colour) to a small area north of the St. Lawrence.

Society of Anti-Gallicans, *A new and accurate map of the English Empire in North America: Representing their rightful claim as confirm'd by charters, and the formal surrender of their Indian friends, likewise the encroachments of the French, with the several forts they have unjustly erected therein*, 1755, NMC 6651

A MAP
Exhibiting all the New Discoveries
in the Interior Parts of
NORTH AMERICA,
Inscribed by Permission
To the Honorable Governor and Company of Adventurers of England
TRADING INTO HUDSONS BAY.
In testimony of their liberal Communications
To their most Obedient
and very Humble Servant. A. Arrowsmith.

Nº 24 Rathbone Place.
January 1.ˢᵗ 1795.
Additions to 1811

CHAPTER 1

Passage to the Orient

Standing with his companions on a rock outcrop near Echo Harbour on the Pacific Coast, not far from present-day Ocean Falls, British Columbia, Alexander Mackenzie used a pomade of vermilion face paint and melted bear grease to inscribe a brief

Later in life Mackenzie admitted that "the practicability of penetrating across the continent" was the "favourite project of my own ambition." His historic quest for a Northwest Passage resulted in two expeditions of discovery by the age of 29. The first (1789) led him down the Mackenzie River to the Arctic Ocean. The second (1793) took him up the Peace River to the Fraser, then overland to the Bella Coola River and the Pacific Ocean. This last expedition gave him the distinction of having been the first European to cross the continent.

Thomas Lawrence, *Alexandre [sic] Mackenzie*, engraving, ca. 1810, e002139974

memorial to his epic journey across the Great Divide: "Alexander Mackenzie, from Canada, by land, the twenty-second July, one thousand seven hundred and ninety-three." With those words the Scottish-born fur trader effectively closed the books on more than two and a half centuries of speculation about the nature and extent of the North American interior. As the first European to cross the continent—north of the Spanish-held possessions—Mackenzie killed all hope that a quick and easy passage to Asia lay hidden somewhere in the west.

Initially, the great cartographers of Europe predicted that the newly discovered western hemisphere was nothing more than a narrow strip of land, just a short distance from the treasures of the Far East. Some of their maps even suggested they would eventually find the northern regions of the two continents attached and sharing the same river systems and mountain ranges.

But by the opening decades of the seventeenth century, Europe was beginning to realize just how much farther the Orient actually was from the New World. In desperation, European adventurers began looking for a water route that would somehow lead them around or through this great land mass. Some modern-day historians wonder whether the explorers of the New World really believed an all-water passage through North America to India and Cathay was possible. One thing is certain, however. The spices and silks of the Orient guaranteed more lucrative returns than anything North America could provide, including its fledgling fur trade. If the explorers had not used the Northwest Passage and its lure of great profits as an incentive for the exploration of the North American

La table des Isles neufues, lefquelles on appelle isles d'occident & d'Indie pour diuers regardz.

When Sebastian Münster first published this map of the Americas in 1540, European cartographers hoped that the western hemisphere would be nothing more than a narrow strip of land, just a short journey from the spices and silks of Cathay (now China). Japan appears as the island "Zipangri."

Sebastian Münster, *La table des isles neufues lesquelles on appelle isles d'occident & d'Indie pour diuers regardz*, Basel?, Heinrich Petre?, 1552, NMC 21090

interior, then in all probability they would not have received any financial backing for their expeditions into *terra incognita*—unknown lands.

Given these financial realities, it is not surprising that the great explorer and statesman Samuel de Champlain tried to make the most of any natural feature that hinted of a marine passage through the continent. As early as 1601, he was promising his benefactors in France that the St. Lawrence River was their dream highway to the treasures of the Far East. "By this [route]," Champlain hastily concluded, "we will be able to go to Cathay … We will be able to make the voyage in one month or six weeks without any difficulty."

In order to satisfy the direction given to him by his backers—and possibly his own curiosity—Champlain usually tried to question any Aboriginal groups he encountered about the country that lay beyond his own sphere of knowledge. "I gave a hatchet to their chief," records Champlain about one meeting he had with some Ottawas at the mouth of the French River on Georgian Bay, "who was happy and pleased with it as if I had made him a rich gift and, entering into conversation with him, I asked him about his country, which he drew for me with charcoal on a piece of tree bark."

In a chance meeting with some First Nations while exploring the upper Great Lakes, Champlain heard of a great inland sea. "These said savages from the north say that they are in sight of a sea which is salt. I hold that if this be so it is some gulf of this our sea which

Summarizing a lifetime commitment to the exploration of the North American continent, Champlain's magnificent 1632 map of New France was his greatest cartographic achievement. Later European mapmakers used it as a prototype for almost a century. His *Mer du Nort Glacialle* is shown as a large inlet on the west side of Hudson Bay.

Samuel de Champlain, *Carte de la nouvelle france: augmentée depuis la derniere servant a la navigation faicte en son vray meridien,* ... Faicte l'an 1632 par le sieur Champlain. From *Les voyages de la Nouvelle France occidentale, dicte Canada* ..., 1632, NMC 51970

Like many of his contemporaries, the French cartographer Alexis Hubert Jaillot thought that North America was simply an extension of Asia. In this 1694 world map he connected the two continents at what we now know as the Bering Strait. Champlain's mythical *Mer Glacialle* lies northwest of Hudson Bay.

Alexis Hubert Jaillot, *Nova orbis tabula, ad usum serenissimi Burgundiæ Ducis,* Paris, 1694, NMC 12939

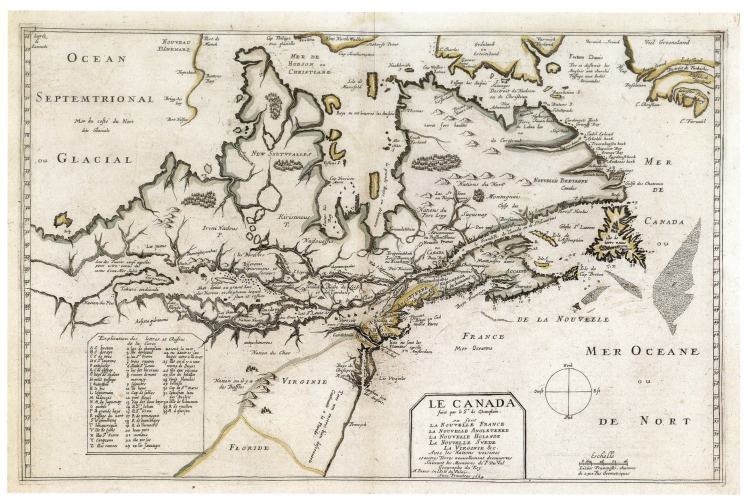

Although exploration west of the Great Lakes came to a standstill the century after Champlain's death in 1635, mapmakers were no less optimistic that a Northwest Passage would be found somewhere to the west. In Pierre Du Val's 1664 map of New France, Champlain's mythical *Ocean Glacial* covers most of the Canadian prairies under a sea.

Pierre Du Val, *Le Canada faict par le Sr. de Champlain, ou, sont, La Nouvelle France, la Nouvelle Angleterre, la Nouvelle Holande, la Nouvelle Svede, la Virginie &c. avec les Nations voisines et autres Terres nouvellement decouvertes suivant les memoires de P. du Val, géographe du Roy*, Paris, 1664, NMC 8757

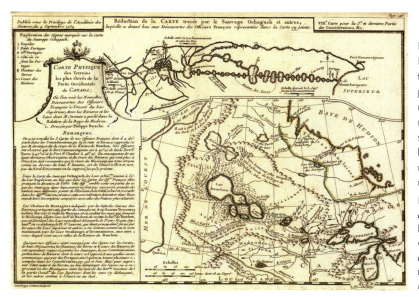

Philippe Buache's 1754 map of the region west of Lake Superior attempts to show the two conflicting theories about the nature of the country that were prevalent at the time: the east-west flow of rivers mapped by La Vérendrye (top map) and the northwest-southeast flow (bottom map) reported by some French adventurers. Together, the two maps hint at some of the theoretical problems European mapmakers faced as they tried to unravel, sight unseen, the mysteries of Canada's western landscape.

Philippe Buache, *Carte physique des terreins les plus élevés de la partie occidentale du Canada ...*, Paris, Publiée sous le privilege de l'Academie des sciences, 1754. From *Considerations géographiques et physiques sur les nouvelles découvertes au nord de la grande mer, appelée vulgairement la mer du Sud: avec les cartes qui y sont relatives*, 1754, NMC 13295

In this map of North America, the French cartographer Bellin extended the Lake of the Woods and its drainage system to the Pacific coast. At the time, it was extremely difficult for navigators to estimate their longitude (east-west distance from a meridian). Accurate longitudes can be obtained by measuring the angular displacement of the moon from other celestial bodies, but the method requires detailed tables that give precise measurements on the motion of the moon. Although the *Connaissance des temps* contained tables as early as 1679 to assist in calculating lunar distances, the first complete set of practical tables was not available until the publication of the *Nautical Almanac* in 1766, almost a quarter century after Bellin's map.

Jacques Nicolas Bellin, *Carte de l'Amerique septentrionale pour servir à l'Histoire de la Nouvelle France*, Paris, 1743. From *Histoire et description générale de la Nouvelle France: avec le journal historique d'un voyage fait par ordre du roi dans l'Amérique septentrionale*, NMC 98179

overflows in the north in the midst of our continent [Hudson Bay]." His mythical sea—which he labelled "Ocean Glacial" or "Mer du Nort Glaciale"—was thought to extend south and west of Hudson Bay, covering much of what we now know as the Prairie provinces under its icy waters. Champlain first incorporated the Ocean Glacial into his manuscript map of New France in 1616 and again in 1632. So strong was the French hope in its existence that it remained a part of their North American cartographic tradition for the next 150 years.

After Champlain's death from a paralytic stroke on Christmas Day 1635, the French dream of reaching the western ocean fell victim to administrative restraint and poor judgment. The search for the western ocean did not pick up again until the late 1720s, when Pierre Gaultier de la Vérendrye took command of the French settlements at Kaministikwia, Nipigon, and Michipicoten in the upper Great Lakes. His interest focused on the string of lakes and rivers west of Lake Superior, which we now know as the Lake of the Woods system. He built a series of trading posts along this system

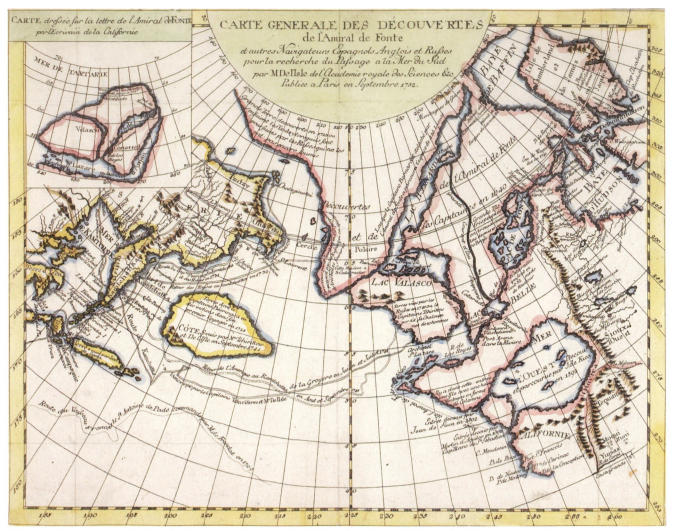

After spending 21 years at the Russian Academy of Sciences in St. Petersburg, Joseph Nicolas de L'Isle, a cartographer of considerable note, returned to Paris to join the mapping business of his brother-in-law, Philippe Buache. He brought with him a description of the de Fonte voyage, which Buache incorporated into his 1752 map of North America. Others began copying it, each adding his own interpretation, with Robert de Vaugondy publishing this elaborate "masterpiece." California is shown in the lower right corner, Greenland in the upper right.

Didier Robert de Vaugondy, *Carte generale des découvertes de l'Amiral de Fonte: et autres Navigateurs Espagnols, Anglois et Russes pour la recherche du Paysage a la Mer du Sud par M.De L'Isle de l'Academie royale des Sciences &c*, Paris, 1752. From *Encyclopédie, ou, Dictionnaire des sciences, des arts et des métiers*, 1751–65, NMC 6913

from Fort Saint Pierre (Rainy Lake, Ontario) to Fort Rouge (Winnipeg). Aboriginal accounts in this area told him of a river that flowed straight west to a great inland lake called Lac Ouinipique (Stinking Water). Another river was said to continue from the outlet to this lake for some ten days and then empty into a large ocean. La Vérendrye concluded that this ocean was the Pacific, and that the river's final destination would be somewhere north of California.

The maps La Vérendrye sent back to colonial administrators in Québec received a great deal of attention. They were eventually re-drafted and transferred to Paris, where several contemporary mapmakers made

use of them in their compilations of French possessions in the New World. Jacques Nicolas Bellin was among the first to take advantage of this new information, and his 1743 map of North America clearly shows a great river stretching westward across the heart of the Canadian Prairies.

La Vérendrye and his four sons spent the rest of their lives searching, without success, for the mythical "Lac Ouinipique." Because of the trust they inspired in the Plains Indians, two of his sons were able to march across the Prairies in 1743 to within sight of the Rocky Mountains. Their explorations should have been sufficient to dispel the rumours of a western ocean, but this was not the case. So strong was the belief in the ocean's existence that French cartographers merely relocated it on the other side of the Rockies. "It is precisely on the other side and at the foot of these mountains," argued Father Castel, a Jesuit missionary and scholar based in Paris.

Joseph-Nicolas de L'Isle and Philippe Buache reflected these sentiments in their 1752 map of North America, but they added a new twist. They connected the western sea—which they called the Mer de l'Ouest—to the Pacific through the Strait of Juan de Fuca. Although de L'Isle and Buache managed to place the Lake of the Woods system closer to its actual position, and were among the first of the European cartographers to show Asia and America in their proper relationship, they still thought an inland sea must occupy a large portion of present-day southern British Columbia and the state of Washington.

The de L'Isle and Buache map of the northwest coast was based largely on the published voyages of the Spanish admiral de Fonte, who claimed to have sailed about 2,400 kilometres into the North American interior via the Northwest Passage. The de Fonte account was a complete fabrication, but some eighteenth-century armchair cartographers were so convinced a Northwest Passage would be found they were willing to believe anything that hinted at its existence, even if their only proof lay beyond the bounds of reasonable credibility. Unfortunately, Russian secrecy concerning their trading activities in the north Pacific did nothing

Philippe Buache's 1752 map showing the mythical "Mer de l'ouest" was based on the voyages of the Spanish admiral de Fonte, the descriptions of which were brought to France by his brother-in-law, Joseph-Nicolas de L'Isle. Only once the inland sea was proved not to exist, did Europe's mapmakers realize de Fonte's travels were purely imaginary.

Philippe Buache, *Cartes des nouvelles découvertes au Nord de la Mer du Sud tant à l'Est de la Siberie et du Kamtchatka, qu'à l'Ouest de la Nouvelle France,* Paris, Philippe Buache, [1752], NMC 21056

WOODBLOCK PRINTING

Many of the maps that document Europe's early search for a western sea were produced from woodblock printing. The technique called for planks to be sawn along the length of the tree, exposing the horizontal grain. The map's image was then carved into the well-planed, smooth surface of the plank by cutting away all areas that were not to be printed, leaving a mirror image of the map in relief on the plank's surface.

After the carving was completed the image was transferred to paper, first by rolling black ink across the raised lines of the plank, then by pressing paper against the inked surface. Although simple hand pressure might be adequate for the transfer to take place, the additional pressure of a printing press resulted in a more consistent image.

Before woodblock printing, European mapmakers were obliged to draw each map by hand, a time-consuming and expensive process. The arrival of the printing press effected a momentous change, spawning an information revolution whose impact is difficult to comprehend today. The technology had its drawbacks, however, of which the most restrictive were directly related to physical limitations in the wood plank. For example, because of the tendency of wood to be affected by humidity, the best prints were made from smaller planks, which were less likely to warp and split, but still shrank and

The German cartographer and cosmographer Sebastian Münster first published his famous woodblock map of the Americas in his *Geographia Universalis* (1540) and, later, in his *Cosmographia*. The latter proved so popular that the publication went into 40 editions. Before woodblock printing, such wide dissemination would have been impossible.

Sebastian Münster, *Sei Libri della Cosmografia uniuersale ...*, Basilia, Stampato a spese di Henrigo Pietro, 1558

Published in Venice in 1556, Giovanni Battista Ramusio's woodblock map covers the Atlantic coast from Narragansett Bay to Labrador, including Cape Breton and Newfoundland (the collection of islands to the east of the main land). The stippled string running across the bottom and up the right side of the map is thought to represent either the Grand Banks or the Gulf Stream.

Giacomo Gastaldi and Giovanni Battista Ramusio, *La Nuova Francia*, Venetia, Nella stamperia de' Giunti, 1556. From *Terzo volume delle nauigationi et viaggi: nel quale si contengono le nauigationi al mondo nuouo, alli antichi incognito, fatte da don Christoforo Colombo genouese, che fu il primo à scoprirlo à i Re Catholici...*, 1556, NMC 52408

expanded from one day to the next affecting the accuracy of the map's scale. Larger maps were possible, but only if mapmakers were willing to divide the map into sections and print each section from a separate block.

With woodblock printing it was also impossible for mapmakers to achieve tonal effects or shading. They could break up the monotony of large white surfaces by adding lines or stipples, but again, because of the nature of wood, there were limitations on how close together these features could be placed.

Despite its disadvantages, woodblock printing remained popular from the first discoveries of the western hemisphere through to the seventeenth century. When coupled with the printing press, the technology allowed Europe to disseminate a graphic vision of the world in unprecedented numbers and at unparalleled speeds. For the first time in human history, people in different cities, different regions, and different kingdoms had access to the same geographical information.

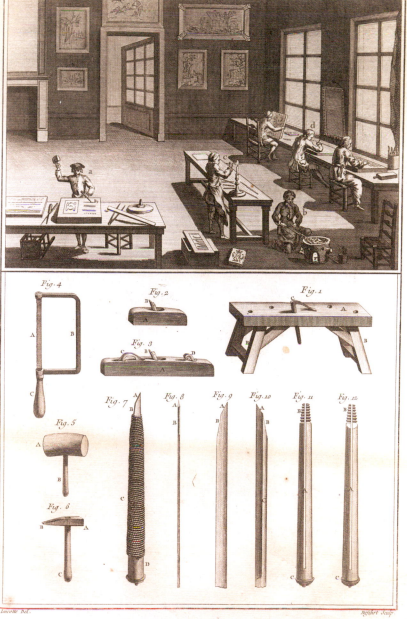

An 18th-century engraving showing some of the tools and tasks involved in preparing a wood plank for printing. Usually, the carving was done with a bevel-edged knife with a triangular point. The knife was used to remove thin slivers of wood from the edges of the image; broader chisels and gouges were used to remove the material between the knifed outlines. Woodblock printing required that all lines and lettering be much heavier than other later technologies. It also limited the amount of information a mapmaker could put on a map and its placement. In the early 1800s, printers found that many of the problems associated with wood planks (shrinking, warping, and cutting across the grain) could be eliminated by carving the image into the end grain of a block of hardwood instead of into the side. However, the wood engraving process was limited primarily to smaller illustrations, such as those used in books and magazines, rather than to larger maps.

Jean Le Rond d'Alembert and Denis Diderot, *Encyclopédie, ou, Dictionnaire des sciences, des arts et des métiers ...*, Paris, Chez Briasson ..., David l'aîné ..., Le Breton ..., Durand ..., 1751–65, 17 vols.

Gravure en Bois, Outils.

The Arrowsmith firm in London had access to the records of the Hudson's Bay Company, enabling it to publish this wall map showing the latest discoveries. First printed in 1795, the map was updated regularly to 1850. In this 1802 edition, Samuel Hearne's trek down the Coppermine River to the polar sea is plotted (top, left of centre), as well as the explorations by Alexander Mackenzie down the Mackenzie River and across the western cordillera to the Pacific. The Lewis and Clarke expedition to the Pacific (1803–6) was incorporated into later editions of the map.

Aaron Arrowsmith, *A map exhibiting all the new discoveries in the interior parts of North America*, London, A. Arrowsmith, 1802, NMC 19687

to dispel these rumours about the Mer de l'Ouest. If anything, this secrecy served only to fuel European speculation that the Northwest Passage really existed.

A decade after the publication of the de L'Isle and Buache map, the Treaty of Paris (1763) left France without any significant possessions in North America. Their interest in the western sea now relegated to a purely scientific exercise, it fell to the British to finish the search. The de L'Isle and Buache map had sparked their interest, and a princely reward of £20,000 was posted for the first ships of the Royal Navy to confirm the

existence of the sea and its relationship to the Pacific and Arctic coasts. Over the next quarter century, Britain mounted a two-pronged assault on the two hypothetical exits to the inland sea. With the cooperation of the Hudson's Bay Company, an overland expedition from Fort Prince of Wales (Churchill, Manitoba) attempted to approach the northern exit, while navigators along the Pacific coast approached the western exit.

The northern expeditions were led by Samuel Hearne. Setting out on foot and carrying a "large skin of Parchment that contained twelve degrees of latitude

North and thirty degrees of Longitude West of Churchill Factory," on which he had "sketched all the West Coast of the Bay … but left the interior parts blank, to be filled up during my journey," he reached Coronation Gulf in the winter of 1772–73. Much to his dismay, he found no hint of the "Ocean Glacial" originally proposed by Champlain more than 150 years earlier.

The second assault was mounted by Captain James Cook just three years later and concentrated on the region around the Strait of Juan de Fuca and northward. It was this approach that the French watched closely even though they were now well out of the picture. "We believe that Mr. Cook has taken on a third voyage and that the passage is the objective of his research," wrote one leading French cartographer. "[T]his great man of the sea will perhaps be the discoverer of the Northwest Passage linking the Pacific and Hudson Bay … my opinion on this subject is not only founded on the surveys made by … Juan de Fuca … but on physical analogies with other parts of the globe: all large continental coasts are broken somewhere between their mid regions and the North … But above California our maps show a continuous land … such a continuity, without bays or rivers, is contrary to nature."

Cook was no more successful than Hearne, leaving one contemporary newspaper account to suggest that "the most sanguine, theoretical or practical navigators will give up, probably for ever, all hopes of finding out a passage." But such was not the case. Old theories die hard, especially among people who have a vested interest in them, and despite Cook's findings, some cartographers still insisted that a western sea was a possibility. "With regard to the Northwest Passage," inscribed the internationally respected cartographer Buache de la Neuville on a map published in 1781, "although it has not been discovered, even after so much research, it is likely that it exists."

Well after the Cook voyages, rumours continued to circulate around Britain that Spanish and American traders had sailed to the Mer de l'Ouest through an unknown route off the Strait of Juan de Fuca. The Royal Navy wisely decided that the only way to dispel these rumours was to mount a second expedition to the Pacific coast. George Vancouver, who had accompanied Cook on earlier voyages to the Pacific, was instructed to take charge, having been ordered not "to pursue any inlet or river further than it [appeared] to be navigable by vessel of such burthen as might safely navigate the pacific ocean." Although Vancouver took his instructions seriously, he was convinced even before his departure that he would not succeed. He was sure that the reports reaching Britain were invented so that other countries could take credit if the Northwest Passage should ever be discovered and mapped. In other words, they were trying to guarantee their right of access to what would surely be the super highway of the century.

Despite Vancouver's meticulous charting of west coast waters—his maps were still in use a century after his death—geographers back in Europe would have no doubt found reasons to continue their belief in the passage. Vancouver's discoveries could always be faulted because the inlets and rivers along the Pacific coast were simply too numerous for one person to search in a lifetime.

The final job of demystifying the Northwest Passage's existence, therefore, rested with Alexander Mackenzie. Setting out by canoe from Fort Chipewyan on Lake Athabasca, he approached the western sea from the east by marching through the heart of the region where it was rumoured to exist. His voyage from Lake Athabasca took him up the Peace River, across the Great Divide to the head waters of the Fraser River, down river, overland to the headwaters of the Bella Coola River, then down river to the Pacific coast. On his arrival at Echo Harbour, he heard from the local inhabitants that Vancouver had passed by just six weeks earlier.

Having marched through the heart of the territory where European maps portrayed a great inland sea— and without finding a single hint of its existence— Mackenzie unknowingly introduced a new reality to the North American interior. As long as there was the possibility of finding a route to the Orient, European mapmakers would never be able to accept the western interior on its own terms. The mythical western sea symbolized Europe's hope of enhancing its narrow economic self-interests. Only after the Passage was totally demystified would it be possible for these intruders to turn their attention to the western landscape and begin mapping it for its own secret wealth. ❧

PLAN of the
HABITED Part
PROVINCE
QUEBEC.

by James Peachey, Ens.n 60 Reg.t

CHAPTER 2

General Murray Maps the St. Lawrence

WITH THE FINAL SURRENDER of the French army at Montréal on September 8, 1760, Maj. Gen. Jeffrey Amherst, the commander in chief of His Majesty's forces in North America, found himself in a difficult situation. The British army now had complete control of the entire St. Lawrence valley—an area encompassing some 65,000 *Canadiens*—yet it knew nothing about the region. While it may have been possible for Amherst to wrestle the colony from the French with little geographical intelligence, he needed to be better informed if the new military regime was actually to control it. What General Amherst needed was a map that gave the occupying British army information on the relative locations of towns and parish boundaries, as well as vital strategic information on the roads, rivers, and fords, the possibilities for campsites and provisioning of troops, terrain features, and defensible positions.

With the relaxation in military pressures in New France and while waiting the end of the Seven Years War in Europe, Gen. James Murray—one of three military governors the British placed in charge of the colony—turned his attention to this lack of information. Lieut. John Montresor, a young engineer with considerable mapmaking skills, had managed to survey the upper part of the St. Lawrence River to Montréal Island before the river froze in the late fall of 1760. Murray was quick to realize that if he extended this work to cover all of the St. Lawrence and secondary rivers—the Richelieu, the Chaudière, and the Saint

John—the British would not only have a good idea of the settled parts of their new possession but also a sense of how these river highways connected to the British colonies in New England.

The water routes to New England were particularly worrisome to General Amherst. Earlier in the year, one

Gen. James Murray initiated his project of mapping the St. Lawrence so that the British army would never again "be at a loss how to attack and conquer this country in one campaign." As a military commander, Murray was easy to anger and excite but he was also widely recognized for his courage. James Wolfe held him in high regard and personally selected him as his junior officer in the siege of Québec.

Unknown artist, *James Murray*, oil on canvas, ca. 1770, C 002834

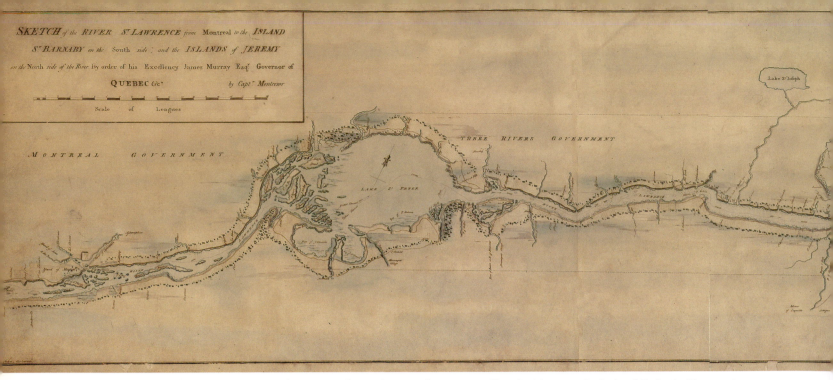

SKETCH of the RIVER S.t LAWRENCE from Montreal to the ISLAND
S.t BARNABY on the South side; and the ISLANDS of JEREMY
on the North side of the River by order of his Excellency James Murray Esq.r Governor of
QUEBEC &c.a   by Capt.n Montresor

Scale   of   Leagues

MONTREAL   GOVERNMENT

THREE   RIVERS   GOVERNMENT

LAKE S.t PETER

Lake S.t Joseph

The sketch map of the St. Lawrence River between Québec and Montréal was surveyed by John Montresor in the late fall of 1760. The map is oriented to magnetic north, and reflects Montresor's reliance on compass bearings. His survey became the basis for Murray's monumental map of the St. Lawrence the following year. Unfortunately, Montresor's survey notes covering the region between Québec and the Kennebec River, Maine, later fell into the hands of Benedict Arnold, who used them as a guide for his assault on Québec in 1775. Montresor disliked Murray intensely and referred to him as a madman. Their animosity further soured relations in the drafting rooms of Québec.

John Montresor, *Sketch of the River St. Lawrence from Montreal to the Island of St. Barnaby on the south side, and the Islands of Jeremy on the north side of the river*, Québec, 1760, NMC 16842

scouting party that had been sent to find such a connection along the Chaudière and Androscoggin rivers came close to disaster. As Amherst himself put it, because "they knew so little of the distance they were to march," the expedition fell short of provisions and had had to exist on a diet of bullet pouches, shoes, and boiled twigs for the last thirteen days. At the very least the episode underlined the limited knowledge the British command had of their surroundings and how dangerous this limitation might prove to their flimsy hold on the French colony.

No doubt Murray also had some practical considerations in mind when he proposed the survey. Because they were an army of occupation, there was a temporary surplus of engineers in the colony; the survey would help keep them busy while compiling some useful information. A bored Murray even thought that he might be allowed to help by undertaking a reconnaissance of the Saint John River. "Since you can give me no better employment … this will at least amuse me, and may perhaps be of some use hereafter," he wrote to Amherst. But General Amherst would not hear of it. "Your going across the country down the St. John River … will take you too great a distance from Quebec," complained Amherst in his response. "One is never sure of what may happen." Instead, Murray sent a junior officer, Lieut. Joseph Peach of the 47th

Regiment, to investigate the river. Amherst's concern proved to be well founded. Peach was in the field two months longer than any of the other officers. As an indication of the magnitude of the task, his rough sketches were so hard to decipher they had to be returned for clarification and, in the end, could not be incorporated into any of the British maps.

There was also the possibility that the St. Lawrence colony would be returned to France once the two belligerent nations concluded a treaty (which they eventually did with the signing of the Treaty of Paris on February 10, 1763). In the event of a future campaign against the French in this part of North America, they wished for a thorough knowledge of the countryside. "The whole survey will be finished by August," boasted an optimistic Murray in a letter to Prime Minister William Pitt in London. "Happen what will, we never again can be at a loss how to attack and conquer this country in one campaign." As fate would have it, the colony was not returned and the British continued their surveys, next targeting the channels, anchorages, and shoals in the waters around Newfoundland and Labrador (Michael Lane and James Cook, 1763–75); Nova Scotia (Joseph F.W. Des Barres, 1764–75); and Prince Edward Island, Îles-de-la-Madeleine, and Cape Breton (Samuel Holland, 1764–76). These coastal surveys were eventually published almost in their entirety

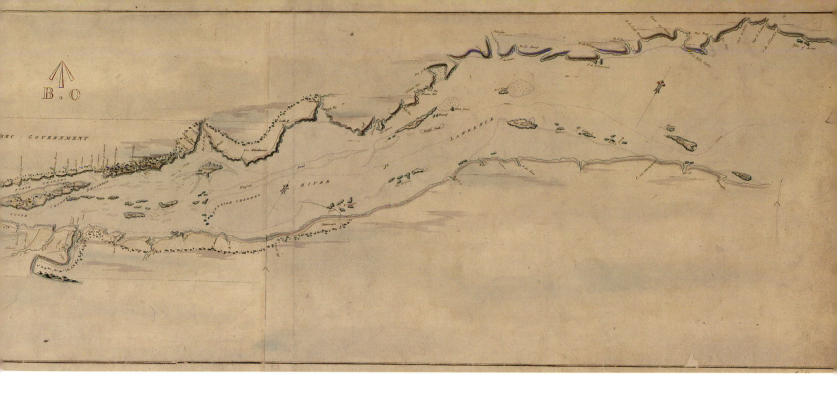

as *The Atlantic Neptune*, easily one of the largest hydrographic publications attempted to date (chapter 7).

Neither Murray nor Amherst gave any direction as to what was needed in the way of a St. Lawrence map, other than their insistence that the work was to be a general reconnaissance and completed quickly (again, because they were expecting the colony to be returned to France). Although there was no consensus about what would be considered the optimal scale for such a military survey, the engineers would have understood that they should focus on the depiction of settlements, roads, streams, and surface cover, especially woodland, swamps, and farmland. This required angles to be taken with an instrument and distances measured with a chain, which would allow for a map capable of guiding commanders to and from the battlefield and finding a favoured position. In other words, it involved route selection, encampment identification, siege craft, and logistical support. Whereas the position and arrangement of hills was important, for example, their height above sea level (as shown on most later-day military maps) was not.

Since there was no time to obtain equipment from England, the surveyors had to rely on the instruments that most engineers in the British army carried on campaign: a circumferentor (a compass with sights mounted on a moveable arc with a spirit level), a plane table, measuring chains in 16-, 22-, 50-, and 100-foot lengths (4.8, 6.7, 15.2, and 30.5 metres) and a quadrant (an instrument for measuring the height of celestial objects in the calculation of latitude). If anything was lacking, they had to improvise. With such equipment, high-level precision was not possible. Because of temperature changes or faulty construction, for example, surveyor's chains might vary by as much as 16.5 centimetres and the circumferentor allowed angle measurements only in one-degree increments (more precise instruments existed, but they were not available to British engineers in Canada). There is even some suggestion that the telescopes on their circumferentors were not equipped with cross hairs (to ensure accurate sightings on distant objects) and were characterized by their users as "common" and "inferior": a miscalculation of five kilometres over a distance of twenty-nine was not uncommon. Although cross hairs had been used in France since 1669, the historian Douglas Marshall found no evidence of them in British military surveys until Samuel Holland's surveys of Lower Canada in 1764.

Field personnel for Murray's project included, from the regular army, Lieutenants Samuel Holland, Joseph Peach, Lewis Fusier, Frederick Haldimand, and Ensign Philip Pittman; and from the Corps of Engineers, Capt. William Spry and Lieut. John Montresor. Despite

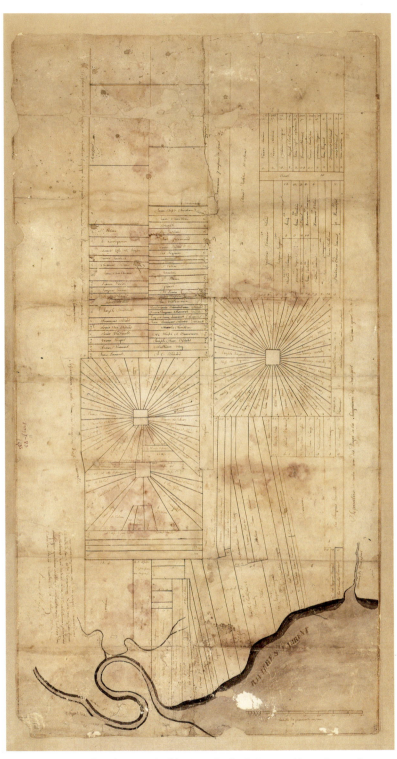

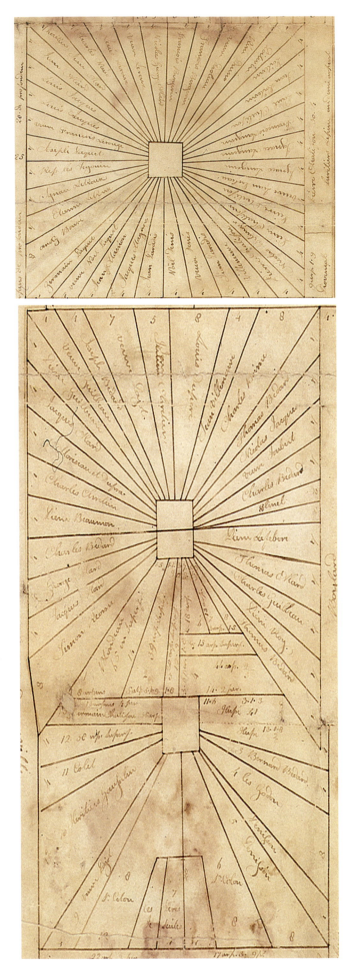

Before the arrival of the British, the Seigneury Notre-Dame-des-Anges was mapped in detail by the French government; it was the exception, however. Most regions of the colony were not described cartographically and, if the British were to administer the colony properly, they would need to undertake their own surveys. Notre-Dame-des-Anges was one of North America's first planned communities. The regions of the seigneury behind the river lots were laid out so as to give each farm equal access to the road network, in the same way that long narrow river lots provided equal access to the St. Lawrence. The map also records the names of the lot owners in the seigneury.

Mtre. Plamondon, *Plan du fief et seigneurie Notre-Dame-des-Anges ou Charlesbourg Royal*, 1754 (1823), NMC 18555

the protests of some officers who resented losing lower ranks to the Ordnance Department, claiming it made them "useless as soldiers," the garrison at Québec also provided recruits to carry out camp duties, transport baggage, and perform manual labour for the survey crews. With this complement from the enlisted ranks, the surveyors probably ran six- or seven-man teams. Since the objective of the St. Lawrence survey was the creation of a handsome and workable map, the Murray team also included three draftsmen: Charles Blaskowitz, Digby Hamilton, and Charles McDonnell.

Murray's selection of Captain Spry was unfortunate. Like many graduates of the engineering academy at Woolwich, Spry could, in Murray's estimation, "neither draw nor survey," but since Spry was a senior engineer, Murray probably felt compelled to place him in charge of the project. However, Murray was luckier in having Samuel Holland and John Montresor on staff. Montresor had acquired survey training at the early age of fourteen from his father—also an engineer in the British army—while both were stationed at Gibraltar. He saw his first action in 1758 at the British siege of the French fortress of Louisbourg on Île Royale (Cape Breton Island). Following his transfer to the Québec theatre of war, he produced a number of plans of areas along the St. Lawrence outside Montréal. This practical experience put him far ahead of his colleagues, and perhaps explains why some of the supervision became Montresor's responsibility, despite his lower rank. Holland, on the other hand, seems to have picked up his surveying skills while serving as an officer in the Dutch artillery. He transferred to the British army in 1756, and by 1758 he too was serving as an assistant engineer in the siege of Louisbourg. He proved his competency by surveying the ground adjacent to the fortress and nearby harbour while under heavy fire. Holland's drawings eventually enabled Brig. Gen. James Wolfe to capture the town.

It seems the survey parties operated independently and made no attempt to connect their survey lines until they met back in the drawing room in Québec during the winter months. The field crews spent most of the spring, summer, and fall of 1761 surveying a one-to-four kilometre strip along both shores of the St. Lawrence corridor from Les Cèdres above Montréal to Île-aux-Coudres below Québec near Baie-

Lieut. John Montresor, one of the principal surveyors of the Murray map, was among the more capable and ambitious officers to serve in Britain's 18th-century North American army. He served under fourteen commanders and participated in more than twenty engagements. "If he had commanded British troops in America," writes the historian Kenneth Roberts in *March to Quebec*, "the American Revolution would have ended in 1776."
Unknown artist, *Captain Sir John Montresor*, watercolour on ivory, ca. 1765, C 133736

Saint-Paul, along with parts of the Richelieu and Chaudière rivers that provided communication with New England. After the survey had begun, Murray decided to enhance the final work by including a census of the region. The result was a detailed description of each parish along with a census that took particular note of the number of men able to bear arms. This last feature set the Murray map apart from all other eighteenth-century British colonial mapping: the Watson-Roy map of Scotland (1747–55), the Vallancey map of Ireland (1778–90), the Holland charts of the North American east coast (1764–75), the De Brahm map of Florida (1765–71), and the Rennell survey of Bengal (1765–77), none of which were to include any observations on the families that the British administrators were expected to govern.

Although there is no record of how the survey was carried out, it is likely that each surveyor kept a field book for measurements and a sketchbook for quick drawings of the countryside. Details of the terrain were either drawn freehand in the sketchbook or were

BRITAIN'S MILITARY ENGINEERS

In the eighteenth-century, British army engineers were generally responsible for the conduct of sieges and for the construction of fortifications, roads, and other military buildings. The work required a reading knowledge of scaled drawings and an ability to make small-area fortification plans and battle maps. It was work that provided little practice for a "national" survey of the size and scope envisioned by Murray for his map of the St. Lawrence. The Royal Military Academy at Woolwich, where most engineers received their training before an assignment overseas, offered some courses in technical drawing, but knowledge of trigonometry and Euclidean geometry—essential for surveying—were taught, not as part of mapmaking, but of fortification and artillery. The cadets were not given a chance even to "acquaint" themselves with the necessary surveying instruments.

James Peachey prepared this cartouche for a map of Quebec that was never published. Rather than fund Peachey's work, Britain's Board of Trade put its financial backing into a series of east coast hydrographic charts prepared by Joseph F.W. Des Barres under the title *The Atlantic Neptune*. The cartouche shows the high degree of artistic skill acquired by some British draftsmen.

James Peachey, *A plan of the inhabited part of the Province of Quebec*, watercolour and ink over graphite on laid paper, ca. 1785, C 150742

Not surprisingly, British engineers in this period made little contribution to the literature on surveying or to the development of surveying instruments. The centres of innovation lay in other parts of Europe. As Britain lacked a repository of topographical information, the teaching staff at Woolwich had to turn to France and Prussia for textbooks. Holland and Montresor, the only two engineers on Murray's team who were capable of bringing his map to fruition, received their academic training outside regular classes at Woolwich. Practical experience was gained on the job at the sieges of Louisbourg, Québec, and Montréal.

Throughout much of the eighteenth century, the Corps of Engineers had a poor reputation in the British Army, its members often feeling disadvantaged when compared to regular line officers. Until 1769, for example, an engineer's widow and orphaned children were denied military pensions; the engineer himself received no pension on retirement. Line officers changed regiments regularly and could rise rapidly in the ranks; engineers were mired in a seniority system.

Although the Corps itself may have had little influence on the British military mapping, it did provide a number of officers who, despite the odds, managed to produce a disproportionate number of cartographic treasures.

As commanding engineer at Québec, Gother Mann undertook this sketch of Lake Huron as part of his inspection of military posts on Britain's western frontier some two and a half decades after the Murray map. A graduate of the Royal Military Academy at Woolwich, Mann was one of the few engineers in North America capable of undertaking a survey of this magnitude. The plan was used for years as a source for general maps of the lake.

Gother Mann, *Sketch of Lake Huron*, Québec, 1788, NMC 18558

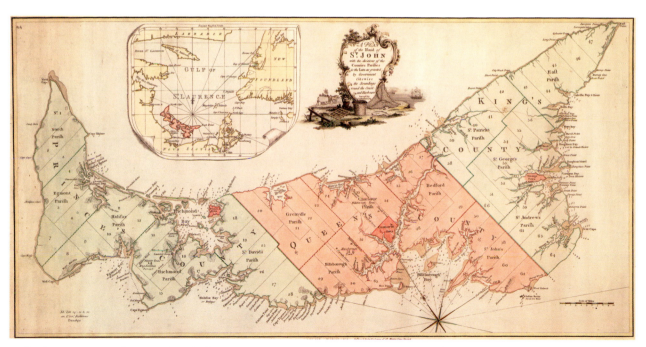

Holland's 1765 survey of Prince Edward Island (known at the time as St. John Island and as Île Saint-Jean under the French) was a prelude to Britain's offering the French settlement to land speculators, military officers, and other friends as a "spoil of war." Since the British government did not have its own publishing house, many engineers submitted their manuscript maps to commercial printers, in this case Jefferys and Faden of London, so that they might have a wider distribution.

Samuel Holland, *A plan of the Island of St. John with the divisions of the Counties Parishes & the lots as granted by Government likewise the soundings round the coast and harbours*, London, Jefferys and Faden, 1765, NMC 23350

"Parish of Varrene's or St. Anne
This Parish originally extended 20 arpents [about 1.2 km] front upon a depth of one league and a half [about 6 to 11 km]. It has been since considerably enlarged at different times. It was granted in 1661 to Mons'. Varrene then Lieutenant of the Regiment Carignan and Governor of Trois Riveres, it is now parcelled out into several small Seignories who have possessed the same by right of inheritance, or purchase; the principal ones are Madame de Varrene, Mons'. Sumandre and Mons'. Louis Minoisboits; the Isle St. Lavoier is a dependence of this Parish; it is not inhabited, being overflowed every Spring; the Island of Therese is another dependence it belongs to Monsieur Langloisiere. It is entirely cleared and has 19 families settled upon it; the Isle Blegard is likewise a dependence thereof and belongs to two brothers of the name of Bienvence....but it is not inhabited upon. Account of its being overflowed every Spring. Towards the center of this parish and about eight arpents distance from the waterside are some salt pans where the French made some salt in their distress just before the surrender of the country, but this was attended with so much difficulty it is now entirely dropped. Families 190, Men able to bear Arms 273."

Charles Blaskowitz, Lewis Fusier, Frederick Haldimand, et al., *General James Murray's map of the St. Lawrence*, Québec, 1761, detail of the St. Anne sheet, NMC 135047

plotted to scale on a plane table. Once in the drafting room, skilful brushwork would hide oversights in the field. There is some evidence that maps were valued more for their appearance than for their accuracy.

Although the fieldwork had progressed without incident and was completed only about four months behind schedule, the same could not be said about the situation in the drafting room. Within weeks of setting up headquarters in Québec, the officers began making petty accusations against each other. The engineers already felt they had little opportunity for promotion in the British army and consequently were particularly careful not to miss any opportunity for recognition from senior command. There was also the possibility of financial gain for whoever had access to the hand-drawn originals. Almost any publisher in England would have been very pleased to turn these into printed versions, since the public's interest in the Americas guaranteed lucrative sales.

Problems began when the officers complained to Murray that Captain Spry made them observe a strict schedule, treating them "like school boys." They also resented his inability to either draw or survey. Their complaints left Murray no choice but to remove Spry and replace him with Montresor. Spry did not take kindly to this loss of command and responded by writing both General Gage and the Ordnance Board, falsely intimating that Murray had no intention of sending them their copies of the map. Murray denied the charge and accused the engineers of trying to "assume all the credit for themselves," which of course was Murray's hope. Spry's letters had the desired effect, and Murray felt compelled to reinstate him.

The situation really started to get out of control, however, when Montresor erased Holland's signature from a copy of the map that was being prepared for shipment. "[As] you can easily imagine," admitted Murray, "Montresor is not held in the greatest esteem here." Later, Murray himself suffered an embarrassing blow to his reputation. Someone in the drafting room purposely sent a copy of the map intended for Prime Minister Pitt to King George III. Pitt's copy had a lavish, hand-drawn dedication to the prime minister that the displeased king called an "unnecessary ornament." By the fall of 1762, morale was so low that productivity had fallen far behind schedule. General Gage and the

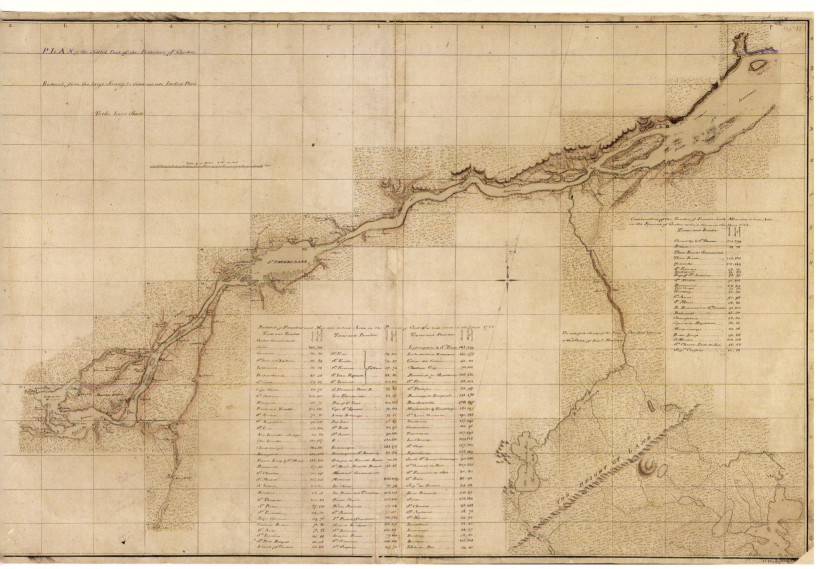

The Murray index shows the extent of the map's coverage along the St. Lawrence, Richelieu, and Chaudière rivers and summarizes the census data collected on each parish by British officers.

Charles Blaskowitz, Lewis Fusier, Frederick Haldimand, et al., *General James Murray's map of the St. Lawrence*, Québec, 1761, index sheet, NMC 135035

Colouring for the Québec sheet of the Murray map followed prevailing 18th-century European conventions: red parallel lines for cultivated fields and roads, blue for water, and grey for relief, with more intense black strokes for steep slopes.

Charles Blaskowitz, Lewis Fusier, Frederick Haldimand, et al., *General James Murray's map of the St. Lawrence,* Québec, 1761, Québec sheet, NMC 135067

Board of Ordnance were each to receive copies, but the draftsmen had returned to their home regiments. Eventually Holland's draftsmen, who were operating under the aegis of the Board of Trade on east coast charts, stepped in and finished the work.

When completed, the Murray map measured an impressive 13.7 by 11 metres and was produced in two formats: one divided the mapped area into four sections, the other into forty-four. The larger maps were sent to each of the heads of state—the king and prime minister—and were perfect for display as war trophies in a grand hall. The more practical smaller maps were obviously designed for use by colonial administrators and were sent to Gage and Amherst; Murray also kept one of the smaller formats for himself in Québec.

The Murray map was one of the biggest and most difficult mapping projects to be undertaken by the British military in North America. Although their survey equipment may have been substandard, the landscape unfamiliar, and the climate unforgiving, the most formidable challenge faced by British engineer-surveyors was one of their own making. It was a product of their seeming inability to organize themselves and properly reward service above and beyond the call of duty. It was found, not at some isolated location far from civilized society, but in the drafting room of their Québec headquarters. Despite the odds, British surveyors were still able to produce a map that would stand the test of time for almost a full century. ❧

A detail view of Murray's mapping of the Montréal area.

Charles Blaskowitz, Lewis Fusier, Frederick Haldimand, et al., *General James Murray's map of the St. Lawrence*, Québec, 1761, detail of the Montréal sheet, NMC 135042

CANADA WEST

ISSUED BY AUTHORITY
OF THE
Honourable
ROBERT FORKE
MINISTER OF
IMMIGRATION
AND COLONIZATION

CANADA — *The New Homeland*

CHAPTER 3

"Free Farms for the Million"

Canada had a problem. It had gone to considerable lengths to acquire the vast western lands of the Hudson's Bay Company, had put together a mounted police force that successfully relegated the region's original inhabitants to reservations, spent two decades carefully subdividing the land into a giant checkerboard of 160-acre homesteads, then paid a king's ransom to service these lands with a transcontinental railway. Yet despite the elaborate preparations and the offer of a free homestead, the expected wave of settlers from Britain and Europe never materialized. In fact, it had been a dismal failure.

In 1881, census-takers counted some 10,000 farms in Canada's great North-West Territories—an area encompassing much of today's Prairie provinces and measuring about two and a half times the size of England. Ten years later, there were still only 31,000 farms in the region. If the settlement were allowed to continue at such an abysmal rate, it would take more than five hundred years to fill the 1.25 million homesteads that surveyors had carved out of the wilderness. Where were the men and women of muscle who would turn this naked landscape into an agricultural paradise?

For more than two centuries, the Hudson's Bay Company had aroused the British imagination with images of the West as an inhospitable wilderness populated by savage "red men." The "great lone land," as one adventurer called it, served the fur-trading interests of the Company well. It kept settlement to a

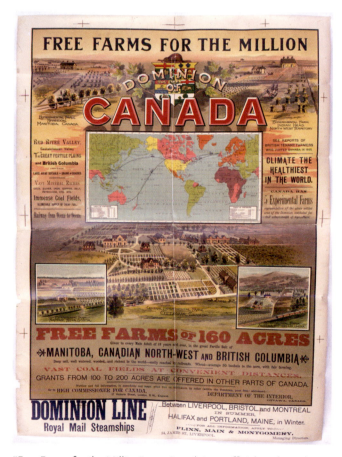

"Free Farms for the Million" was Canada's unofficial credo at the end of the 19th century. Once the agricultural potential of the Prairie West was fully realized, Prime Minister Wilfrid Laurier confidently predicted Canada's population would reach 50 million by 1950.

Canada, Dept. of the Interior, *Free Farms for the Million*, lithograph, Ottawa, Dept. of the Interior, ca. 1890, C 095320

Britain's John Bull and America's Uncle Sam seem quite happy to invest their money in western Canada. Published in a Liberal Party campaign pamphlet, the caricature underlined Laurier's claim that, with proper investment and immigration, the 20th century would belong to Canada.

Liberal Party of Canada, *To Canada, for investment in western Canada*, Ottawa, Liberal Party of Canada, 1904

minimum and preserved the Company's trading monopoly. Although it had been almost thirty years since the Company relinquished its interest in what it called Rupert's Land, the notion of a snow-covered, empty prairie was still deeply entrenched in the collective consciousness of the British public.

For the newly appointed Minister of the Interior, Clifford Sifton, the solution to making Canada better known was obvious. Canada had to create a new image for itself—one better suited to national objectives—and it would have to trumpet this new image through an advertising campaign, the likes of which the world had never before seen. Sifton figured that immigration should be like any other commodity. "Just as soon as you stop advertising," he warned the House of Commons in 1899, "the movement is going to stop."

Officials in the Department of the Interior, where the Immigration Branch was located, had seen how map brochures and pamphlets had changed people's attitudes to the Klondike, turning the Yukon gold rush into a marketing paradise for western communities (chapter 8). Surely an attractive thirty- to forty-page atlas with up-to-date information on the Prairie West would work in Canada's favour. It was one thing to present British families with the "facts" on immigrating, quite another to package these facts so that "the reader becomes impressed by them," as one of his senior immigration officials in the United States candidly pointed out to Sifton in a 1903 report. Prime Minister Laurier's cabinet could not have agreed more. Over the next few years it progressively increased Sifton's advertising budget by some 400 percent, until it had reached a staggering $4 million in 1905. The increase gave Sifton the added flexibility he craved, allowing him to boost the production of immigration literature to more than one million pieces a year.

The Canadian government's emigration office at Trafalgar Square was a "parlour of seduction." In 1911 more than a million people inspected the window displays and picked up information on Canada. In that same year, 123,013 people immigrated to Canada from the United Kingdom, another 66,620 emigrated from continental Europe, and 121,451 arrived from the United States.

Canada, Immigration Branch, *Messrs. Krag and Dawson entering Canadian Emigration Offices, 11–12 Charing Cross, London, England*, 1911, C 063257

As the owner and publisher of the highly successful *Manitoba Free Press*, Sifton realized that if good writing, attractive photographs, useful maps, and colourful covers were to work together to reaffirm the Prairies as a land of prosperity and happiness, the country would have a powerful tool for enticing farmers in the United States, Britain, and Europe to make Canada their new home. In his mind, it would be easier to attract prospective immigrants if the immense Prairies were reduced to something not all that different from the comfortable landscapes people were being asked to leave behind.

Sifton decided that the atlases should ignore some of the least attractive aspects of western Canada—the cold, the isolation, the dryness—or at least repackage these negatives into something that was more respectable. References to the cold, for example, were replaced by such euphemisms as "bracing," "invigorating," and "healthy," while actual temperatures and rainfall values were given as monthly, seasonal, and

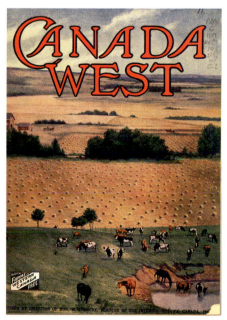

Atlas covers often created a Prairie landscape that was both picturesque and productive. They consistently ignored some of the less attractive aspects of western Canada—the cold, the dryness, and the isolation.

Canada, Dept. of the Interior, *Canada west*, US ed., Ottawa, Dept of the Interior, 1913

W.G. Hunt's sod house (or soddy) outside Lloydminster was typical of many first homes built by Prairie homesteaders. Despite their popularity, images of such rustic homes were never used in Canada's immigration atlases.

John Woodruff, *W.G. Hunt's residence, Lloydminster, Alberta*, ca. 1900–1910, C 014529

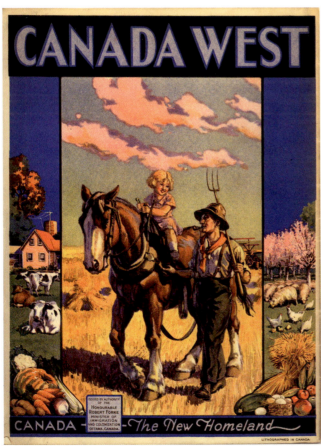

In this atlas cover, the vastness and emptiness of the Prairies is scaled down into a comfortable and inviting family farm scene. Atlas covers often reaffirmed the Prairie West as a land of opportunity by featuring young farm families working together to bring in a bountiful harvest.

Canada, Dept. of Immigration and Colonization, *Canada west: Canada the new homeland*, UK ed., Ottawa, Dept. of Immigration and Colonization, 1928

yearly averages. This last technique was rather clever. Under the guise of providing useful information, the averages hid the daily extremes. Since few immigrants knew the average temperatures of their own country, they would be unable to use the figures as a basis of comparison.

The artwork and photography were carefully selected to support many of the claims made by the writers. Not a single winter scene appears in any of the atlases: it is always summer, and the fields in crop or in the process of being harvested. Farm homes are typically two-story buildings, often brick, and frequently depicted with an automobile or new farm machinery close by. Sometimes an artist carefully added these embellishments to the photograph by hand.

The photographs were taken by some of the best professionals in the business. They are consistently well composed, and while they might have an air of informality about them, internal departmental letters and memoranda indicate that the photographers sometimes went to considerable trouble to compose the shots. Appointments were made at specific farms and businesses on the suggestion of federal authorities and crews worked hard to have machinery, crops, and people in their proper place at the right time. The care taken by the department to commission attractive photographs underlines the importance it placed on this medium.

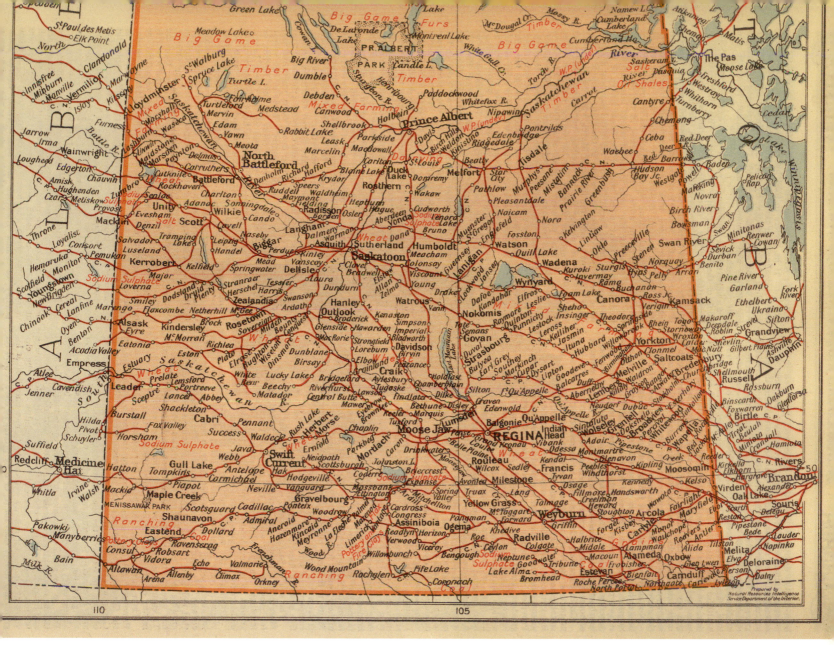

Like other parts of the atlas, the maps refute the isolation of the Prairies. In this map, southern Saskatchewan appears to be well settled with communities that are adequately serviced by rail.

Canada, Dept. of Immigration and Colonization, *Canada west,* US ed., Ottawa, Dept. of Immigration and Colonization, ca. 1928

Without exception, the atlases touted the superlative qualities of the West, using such terms as "unexcelled," "prosperity," and "inexhaustible" throughout every edition when referring to the region's productivity, growing season, and bountiful crops. The colourful, eye-catching covers were especially boastful in this regard, promising not only a productive landscape but a picturesque one as well. The horizon was set high so as to diminish the immensity of the Prairies; everywhere there was lush vegetation and well-established farms. The hardworking but happy farm families are always pictured with bountiful harvests. Nowhere is there any suggestion that homesteaders may have made huge sacrifices in moving to their Prairie farm.

Unlike the painstakingly written text and the carefully composed photographs and covers, the maps

seldom changed from one issue to the next. Each western province was often represented by two, double-page coloured maps of the province's northern and southern regions. The maps were relatively simple, revealing little more cultural information than the surveyed and unsurveyed townships, communities with a post office (with no attempt to indicate the community's size), and railway lines (both completed and projected). Geographical information was limited to the major lakes and rivers. But even at this, most areas of the maps were cluttered with names—English names—that visually gave lie to the emptiness of the West.

The atlases liked to contrast earlier life on the Prairies with the "present" situation. These comparisons helped to impart a sense of progress and success.

GETTING THE MESSAGE OUT

The dream was simple enough. Canada wanted to see the Prairie West become its breadbasket and linked to the world through railways to central Canada. Wheat would flow east to feed Europe, and central Canada would fill the returning freight cars with manufactured goods. Surely, it was reasoned, with the West's expanding population and ever-increasing agricultural bounty, Canada would be able to emulate the success of the United States and perhaps even overtake it. The twentieth century would belong to Canada!

The dream could become reality, however, only if Canada attracted the farmers it needed. "We do not want anything but agricultural labourers and farmers or people who are coming for the purpose of engaging in agriculture," the Hon. Clifford Sifton, federal minister responsible for immigration and western settlement, reminded his deputy minister in a 1904 memorandum. City people, clerks, shopkeepers, and factory workers were not welcome in the new West. "Only farmers need apply" became the underlying message in Canada's immigration literature.

With Sifton's encouragement, Canada used every means at its disposal to distribute immigration literature on "the last best west." It was one thing to design state-of-the-art advertisements, quite another to get them into the proper hands. Since "the agricultural classes" tended to live far from

A successful newspaper publisher and lawyer, Clifford Sifton understood the key to Canada's success lay in the settlement of the West. "The place to which our merchants must look for enlarged markets is Manitoba and the North-West Territories," he once announced to the House of Commons.

William James Topley, *Hon. Clifford Sifton, MP (Brandon, Manitoba)*, 1900, PA 027942

Without exhibition wagons, it would have been impossible for Canada to "so systematically" distribute immigration literature in rural areas. "It is safe to say," reported Canada's immigration agent for Great Britain in 1907, "there is not a highway, nor a village, nor an important by-way of any consequence ... [to which its wagons have] not penetrated with literature, exhibits of Canadian grains and grasses, and drawing the attention of the agricultural people to the claims of Canada."

Canada, Immigration Branch, *Advertising for immigration to Canada...*, 1905, C 075938

normal communication links—railways, lecture halls, and newspapers—the task was not as simple as it might appear.

In Britain, the Canadian government designed exhibition wagons that delivered immigration literature to farmers' doorsteps. Covered from end to end with samples of Prairie produce, the wagons were a natural magnet, attracting large crowds of curious bystanders wherever they went. Once agents had the crowd's attention, it was much easier for them to pitch their story about "the Promised Land" and to distribute "suitable" literature on "Britain's big breadbasket" to farmers.

In the United States, Canada targeted local state fairs. With one out of ten farm families attending these events, the Canadian government spared no expense to set up grand displays from which its immigration atlases and other literature could be distributed. These exhibitions proved so popular that they provoked resentment from American land interests. In 1910 the Canadian government was banned from several state fairs, but because of community protests, was soon reinstated.

Canada wanted farmers from the American Midwest as they were familiar with the hardships Canadian homesteaders faced and had the capital to turn virgin prairie into bountiful farms. This display was just one of several hundred that the Canadian government mounted in any given year for the distribution of immigration atlases and other literature.

Canada, Immigration Branch, *Canadian government exhibition, Oklahoma State Fair*, 1913, C 075991

Modern farm machinery figured prominently in the immigration atlases. If images of older equipment were included, they were often set juxtaposed with the modern to create a sense of prosperity and progress.

Canada, Dept. of Immigration and Colonization, *Canada west*, US ed., Ottawa, Dept. of Immigration and Colonization, 1922, p. 17

The 1910 atlas, for example, contains two maps showing the railways in operation in 1896 and in 1910. The later map also featured future railways—but without giving a timeframe when they would be completed. As a result, western Canada in 1910 appears to be well serviced by rail transport. Comparing the two maps, a viewer cannot help but marvel at how much progress has been made in such a short period.

The atlases also frequently used narration from supposedly real homesteaders, which added a personal note to the official voice espoused by the government's professional writers. Although a systematic survey of every homesteader mentioned in the atlases has never been undertaken, random searches of some of the names in homestead records have failed to turn up any indication that the families existed. The testimonials might have been a clever figment of the Department's imagination—or if genuine, the names may have been changed to protect privacy.

From its first appearance in 1897, the atlas was an immediate hit. An editorial in the *Edmonton Bulletin* declared it "in appearance and in matter … attractive, readable and reliable … The book is worthy of a place in any library and will serve an excellent purpose." Its success encouraged the Department of the Interior (and later the Department of Immigration and Colonization) to turn the atlas into an annual publication. Over the next forty years the atlas would be translated into at least twelve European languages and consume nearly half the Department's advertising budget. At a time when a book was considered a bestseller if it sold 5,000 copies, the Department was ordering print runs of 50,000 to 100,000 at a time, with yearly production figures as high as 400,000 copies. The total print run for all editions of the atlas in all languages is difficult to estimate, but would number in the millions. Despite these impressive production figures, demand always exceeded supply.

Only the English-language edition of the atlas was issued on an annual basis, with separate versions going to the American and British markets. Where the American edition numbered about forty pages, the British edition was usually smaller, around thirty pages. The difference was attributed solely to the higher cost of postage in the United Kingdom. Non-English editions of the atlas were printed only on an occasional basis, and since most European governments took exception to their farmers being actively recruited by the Canadian government, the atlases were primarily distributed to foreign nationals who had arrived in the United States and were looking for land to farm.

To counter British prejudices against the North-West Territories, Deputy Minister James Smart suggested that the Department should make a special effort to target school children. If the government could change their perceptions, perhaps they might think about immigrating to Canada when they got older. With this objective in mind, Smart decided to use the atlas in a nationwide competition, in which British school children would be asked to research and write an essay on Canada. The best essay from each school would be awarded a specially minted bronze medallion, and the grand-prize winner would receive assisted passage for his or her family to the North-West and a free homestead. Although the atlases were handed out to students, Smart instinctively knew that parents would take a look as well. "In this way," mused Smart, "it is thought that the parents of the children will also become interested with a desire to know more of this country"—and perhaps think about immigrating.

The response from British school children was overwhelming. "I am extremely pleased with your free atlas," wrote Edith Beckett to the superintendent of Immigration in London. "It is very much better than I expected it to be. When my mother saw it she said it was very kind of you for sending it free of charge … I have backed it to keep it clean and put it out of the reach of my younger sister." In the first year of the competition, immigration officials in London were inundated with several thousand essays to adjudicate. In subsequent years, student enrolment in the contest grew exponentially.

The effect of the atlas and Canada's advertising campaign on immigration was almost immediately

With their flamboyant and spectacularly colourful covers, it is no wonder Canada's immigration atlases were popular with British school children. "The people like readable facts and maps," observed the Assistant Superintendent of Emigration in 1908, "and … I can conceive of no better value for the expenditure of public funds than … [by] getting [maps and atlases] in the homes of the school children."

Canada, Dept. of Immigration and Colonization, *Canada west*, UK ed., Ottawa, Dept. of Immigration and Colonization, ca. 1923

apparent. In just four years, immigration from the United Kingdom leapt from 17,000 to 87,000 annually. Figures from the United States were equally impressive. Before 1896, American immigration to western Canada was almost non-existent. After Sifton's system of advertising was inaugurated, immigration immediately jumped to 9,000, and by 1906, the yearly figure had reached 57,000. From a near-vacant landscape, the Prairie West expanded to 1.3 million by 1911, and to 2 million by 1921. Winnipeg had been the only Prairie city in 1896, but by 1914 there were twelve. Although motives for immigrating differed widely from family to family—persecution, poverty, lack of opportunity, adventure—within the span of a single generation the West had become "a home for the million." ∾

II
Perfecting
Our Cities

WITH THEIR ARCHITECTURAL SPLENDOURS and scenic wonders, Canada's cities are compelling representatives of the country as a modern industrial state. Calgary with its famous backdrop of mountains, Toronto with its many ethnic faces, and Ottawa with its stately buildings of governance offer powerful statements to the world of the type of society that Canadians envision for themselves. It is not surprising that the maps produced in this country over the last century have called upon the highest skills of the surveyor, artist, and publisher to ennoble our cities and evoke the power and the prestige that Canada seeks on the international stage.

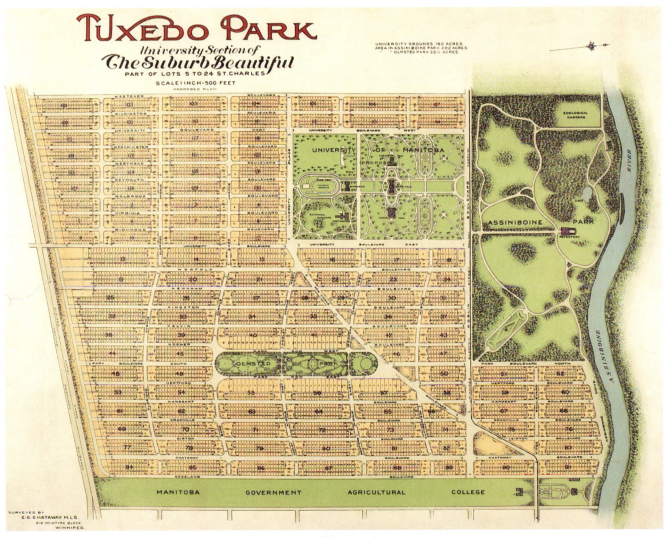

The Tuxedo Park Company used a vertical perspective in this plan of their new housing project in Winnipeg's suburbs. In the tradition of the "city beautiful" movement that swept Canada in the early 20th century, the proposal combined parkland and a university campus to produce the city's "best residential area."

C.C. Chataway, *Tuxedo Park, university section of the suburb beautiful, part of lots 5 to 24 St. Charles*, Winnipeg, Bulman Bros., 1910, NMC 44014

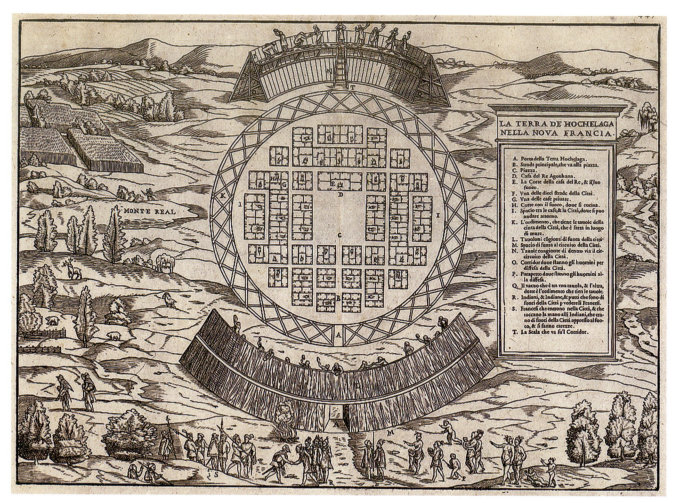

Giovanni Battista Ramusio produced this woodblock print of Hochelaga (Montréal) in 1565. Considered Canada's oldest town plan, the view presented by the Venetian geographer combined two perspectives to convey information on the Iroquoian village. The centre is a vertical plan of the village and the perimeter a bird's-eye perspective of the surrounding landscape.

Giovanni Battista Ramusio, *La terra de Hochelaga nella Nova Francia*, Venetia, Nella stamperia de' Giunti, 1565, NMC 1908

Despite their importance to our nation's sense of self-worth, urban areas have always been somewhat of an enigma for cartographers. Cities have such an incredible array of densely packed features that mapmakers have been forced to develop special ways of showing them. More problematic, however, are architectural monuments, which have challenged mapmakers to devise different methods of expressing the vertical dimension from the more traditional contour lines, hachures, and hill shading generally reserved for rural landscapes.

The three specialized urban maps discussed in this section meet the unique problems that urban areas present in different ways. Vertical plans, which view the city from directly overhead, are very accurate at showing the horizontal placement of a city's features. Since they can be drawn at almost any scale from the detailed to the general, vertical plans, such as those created by Charles E. Goad for fire insurance underwriters, offer the greatest advantage at showing property boundaries, buildings, roads, parks, bridges, utility lines, and transit corridors in relation to each other. Unfortunately, being only two-dimensional, they do not convey a fuller sense of the cityscape.

If a mapmaker is willing to sacrifice the precision of a ground plan for a more pictorial presentation of a city's splendours, there is the bird's-eye view. Seen from an oblique angle high in the sky, the bird's-eye view conveys a sense of the city's monuments, at the same time preserving their horizontal placement—but in perspective rather than at true scale. Cartographers of the late nineteenth century usually reserved bird's-eye views for those instances when the intent was to impress and inspire the viewer with a city's grandeur,

A.H. Hider's late-19th-century lithograph uses a bird's-eye perspective to convey an impression of the vastness of the Gooderham and Worts Toronto distillery while preserving a sense of its overall layout.

A.H. Hider, *Gooderham and Worts Ltd., Toronto Canada, Canadian rye whiskey*, Toronto, Toronto Lithographing Company, ca. 1890s, e000943118

power, and wealth. Although less popular today, they are still used when a cartographer needs to give a general impression of a city's layout while retaining a sense of its profile.

Three-dimensional models, on the other hand, combine the virtues of both plans and bird's-eye views. They preserve the city's physical relief and allow the mapmaker to present the city in all its magnificence while fully respecting its spatial arrangements. Problems of transport and storage limit this form of mapmaking to special situations, however. French colonial administrators produced the first three-dimensional models of Canada's cities: Montréal, Québec, and Louisbourg (unfortunately, none of these models has survived). The English followed in the early nineteenth century with their own model of Québec. It remains the oldest surviving model of a city outside of Europe.

Whatever method is used, one thing is certain: maps provide us with powerful and seductive images of our urban landscape, and are the ideal tools for presenting our cities as "picture perfect" places to live. ∾

PLAN
OF THE FORTIFICATIONS OF
QUEBEC
With the New Works proposed

References

The Black lines unshaded shew the present Works in ruins.
The Black lines shaded yellow, the new permanent Works proposed.
The Black lines, shaded, but not coloured are merely projected. Field
Works, to be raised only as the occasion may require.
The red Figures are the height of ground above the level of the R.S. Lawrence

Scale 200 Feet to an Inch

Spring Tides Rise 22 Feet

B. O

CHAPTER 4

Modelling Québec

"WHEN IT IS CONSIDERED how greatly the safety and preservation of this part of His Majesty's Dominions depend on Quebec," penned Maj. Gen. Gother Mann in a confidential dispatch to the Inspector General of Fortifications in London, "too much attention cannot be paid to promote its strength and security." The thirty-eight-year-old Mann was writing at the close of the eighteenth century, when Britain's relationship with the United States was becoming tense. Although Mann recognized that Canada's harsh winters offered the British some security from a possible America invasion, he readily admitted that something more substantial was required for this military post "of consequence."

Drawing on more than twenty years' experience with coastal defences in England and the Caribbean, Mann identified three major shortcomings to this "Gibraltar of the North," as the British liked to call their Québec fortress: the new stone wall—originally begun by the French more than fifty years earlier—

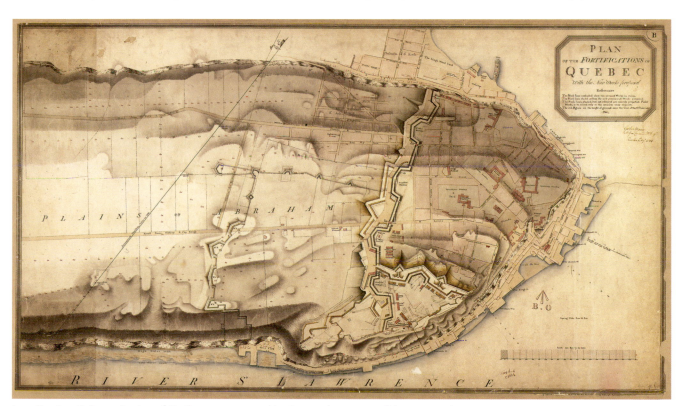

Jean-Baptiste Duberger's plan of Québec was prepared in 1804 to show some of the proposed improvements to the city's fortifications suggested by Maj. Gen. Gother Mann. The detailed map would have provided the measurements Duberger and By needed for their model.

Gother Mann, William Hall, and Jean-Baptiste Duberger, *Plan of the fortifications of Quebec with the new works proposed*, Québec, 1804, NMC 11082

Over the last 200 years, the Québec model has undergone a number of alterations. The first occurred in the early 1900s when the model was repatriated from England by the Dominion Archivist, Arthur Doughty. The 1977 restoration by the Canadian War Museum was the most extensive. It took more than three years, and returned much of the model to its former brilliance.

The Québec model after restoration, ca. 1977,
© Canadian War Museum, 1998, courtesy of the Canadian War Museum

needed to be finished so that it completely encircled the city; a permanent citadel was required on the city's high point if the fortress was to have control of the St. Lawrence River and the major water routes to the colony's western frontier; and some auxiliary defence works were badly needed, both immediately outside the city's west wall and even farther west on the Plains of Abraham. A committee of high-ranking engineers who had authority over such projects for Lower Canada (Britain's North American colony now known as the province of Quebec) had no problem approving Mann's first two recommendations. When it came to the refortification of the city's western approach, however, they flatly turned him down. Ironically, the west wall was the one major weakness in the city's defences that Mann felt would be most vulnerable to an attack, a weakness that the British themselves successfully exploited when they seized Québec from the French during the Seven Years War.

Mann went to considerable trouble to have his defence plans endorsed by the committee. He loaded his reports to England with detailed large-scale maps of the city and its surroundings, "to show as distinctly as possible, the features of the ground." But he now realized that more persuasive techniques were required if his plans for improving Québec's defences were to be fully accepted. With only weeks remaining before his promotion and return to England, Mann took the rather unusual step of organizing the construction of a large-scale model of the city. No doubt his decision to begin such a project, knowing full well he would not be around to see it through to completion, was partially influenced by the "clearer" and "more enlightened conception of the country" three-dimensional models offered over conventional two-dimensional maps and plans.

Mann easily found two bright, enthusiastic members of his staff who were more than willing to undertake his special project: the self-taught surveyor and

PRINCELY TOYS

Although the origins of model making are lost in antiquity, some historians feel it is possibly one of the oldest forms of cartography. One of the first references to "maps in low relieve" was published in 1665 by John Evelyn in the *Philosophical Transactions* of the Royal Society of London. It describes a model of the Isle of Antibe, on the Mediterranean coast of France, as a "new, delightful and most instructive form of Map, or Wooden Country," which can be looked upon "either Horizontally, or sidelong, and it affords a very pleasant object."

It seems model building was introduced to France by Allain Manesson Mallet, the great theoretician on the art of fortification, who got the idea from an unnamed Italian engineer. By the time Mallet published his famous treatise on fortification in 1685, model making was well entrenched in Louis XIV's court. Having instituted a massive fortifications program along the Belgian frontier, the French undertook model making as a matter of course whenever plans were developed for some new defensive works. By the end of the seventeenth century, Louis XIV had amassed some 160 models, Europe's largest collection, which he kept in the main gallery of the Louvre. The king frequently took great pride in showing these symbols of his power and prestige to visiting ambassadors, sovereigns, and military chiefs despite the risk of giving away defence secrets. One visitor, Peter the Great of Russia, is said to have examined these "princely toys" for a long time and to have admired them "with wonder."

Model making was never as popular in England as it was in France. Models of actual English fortifications (such as Québec and Gibraltar) were housed away from the royal palaces. Along with models of hypothetical fortresses and battlefields, they were used primarily for the instruction of junior officers and, as in the case of the Québec model, for military planning.

The famous 17th-century French military theoretician Allain Manesson Mallet presented the world's first complete treatise on model building in his *Les travaux de Mars; ou, L'art de la guerre*.

Allain Manesson Mallet, *Les travaux de Mars; ou, L'art de la guerre, divisé en trois parties...*, Paris, D. Thierry, 1684–85, courtesy of the Thomas Fisher Rare Book Library, University of Toronto

Sébastien Le Prestre de Vauban was considered a brilliant fortifications engineer. The projects he directed for Louis XIV were duplicated in the monarch's model collection and became the focus of military studies in France for more than a century.

Antoine Maurin, *Vauban*, lithograph. From *Iconographie française, ou, Choix de 200 portraits d'hommes et de femmes qui se sont acquis en France ...*, 1840, C 007221

draftsman Jean-Baptiste Duberger and Lieut. John By of the Royal Engineers. Ever fearful that the project might be seen as a frivolous waste of time by his successors to the Québec post, Mann insisted, and Duberger and By readily agreed, that the project would be undertaken at their own expense and "when not occupied in public service."

The two junior staff Mann selected were well qualified for the job. Duberger was working part-time as an assistant draftsman when his maps caught the attention of Mann, who was impressed by their remarkable precision and painstaking attention to detail. Duberger must have been very gifted indeed, since his education was limited to a few years of classical training by priests in one of the colony's seminaries. Most of his skills as a surveyor appear to have been acquired on the job when he accepted a temporary position in 1792 under the Surveyor General of Lower Canada, Samuel Holland. Duberger was thirty-six when Mann gave him his first permanent employment in 1803, a position in the second class of the Corps of Royal Military Surveyors and Draftsmen, Royal Engineers. Duberger was always grateful for the opportunity Mann provided him. "I have never forgot how much I am indebted to you for my happy Situation," he would write years later.

Duberger and By used some unusual techniques to build their model. A wooden base was first built to approximate the topography. A layer of plaster was then applied and moulded to represent changes in the natural landscape more accurately.

Canada, Public Archives of Canada, *J.B. Duberger model of Quebec City seen at the Public Archives of Canada*, ca. 1960s, C 014449

John By, on the other hand, was a graduate of the Royal Military Academy in Woolwich, where he would have used three-dimensional models in his training. He had been busy working on improvements to the defences of Québec when, at the age of twenty-three, he was asked by Mann to take charge of the model project. Given the nature of his posting at Québec, By was probably aware of Mann's proposed changes to the city's west wall and its approaches, and recognized their importance.

Since prior responsibilities prevented Duberger and By from beginning their new assignment until the fall of 1806, they used the approaching long winter months—traditionally a time of considerably reduced activity for the military in British North America—to get their project well underway. They were fortunate in being able to base their model on the notes and sketch maps of earlier detailed surveys Duberger had undertaken for Mann. Once started, the two "worked without relaxation" in order to express "everything whatsoever that can be expressed … All our heights are taken accurately out of which our ground in its irregularity is formed." Even the guns along the walls were represented "according to their different nature … the difference in their Carriages likewise, each of them accompanied with their Piles of Shot, and nothing whatsoever is neglected."

Mann had indicated that he wanted the model built on a scale of 1:900 (1 cm to 9 m). But Duberger and By quickly realized that if they were to proceed with this directive, many of the city's features—especially its fortifications—would be so small that they would be next to impossible to make out. Given "that the intention of Such work is to be turned out to all the advantages of making use of your projects for the defence of this place," By took the advice of Major Robe of the Royal Artillery, who recommended its being done on a scale of 1:300 (1 cm to 3 m).

The model was not built as a single piece, but as nine separate, interlocking tables. Each of the tables differed in size according to the requirements of the total layout and was supported by a frame of white pine. Broad flat pine boards were also used in the tabletop. These boards were planed so that their thickness approximated the ground's topography. Details in the landscape were moulded out of a layer of plaster spread across the surface and painted in appropriate colours to represent vegetation, rock, and water. Buildings were hand-carved out of blocks of wood that were set into the wet plaster at the time of construction or were attached later with animal glue. The architectural features of each structure—doors, windows, and roofs— were not carved into the building blocks but simply painted on the surface. When the model underwent its last major restoration at the Canadian War Museum in 1977, the conservators noted four different carving styles, which suggest that Duberger and By may had have other artisans helping them. Although the model's trees and shrubs have been removed over the years, it is highly likely that these consisted of iron stems wrapped with natural bristle fibres and painted.

The individual tables were built by Duberger at his home, and when each was completed, it was transported down the street to By's lodgings, where the model was being assembled in his front parlour. In order to accommodate the entire model, By eventually tore down the walls separating four rooms.

The work took about a year to finish. When Governor James Henry Craig saw it for the first time, he "expressed himself highly pleased with the correctness of it." But Craig immediately recognized that its merit as a planning tool would be greatly enhanced if the strategic high ground on the Plains of Abraham to the west of the city were added. The addition effectively doubled the size of the model and occupied Duberger and By for another full year. In the end it comprised eighteen sections and measured about 8.2 by 6.1 metres, accounting for more than 445 hectares on the ground.

The final product reflects a curious assortment of defensive works that existed at the time of model construction—such as the temporary citadel and the western ramparts—and others that were only at the planning stage—such as the Martello towers on the Plains of Abraham, whose construction was begun only after the model was completed in 1808. Other elements of Mann's defence plan do not appear at all—the planned permanent stone citadel, for example—despite having been approved by England's committee of engineers as early as 1805.

Parks Canada historian André Charbonneau has weighed this evidence carefully, and suggests that the Québec model was likely requested by Mann as part of

Buade Street © Parks Canada, Québec, courtesy of Parks Canada

Place Royale © Parks Canada, Québec, courtesy of Parks Canada

Collège des Jésuites © Parks Canada, Québec, courtesy of Parks Canada
At least four artisans in the employ of Duberger and By hand-carved the buildings for the Québec model out of small blocks of wood. To make the buildings more realistic, various architectural elements—the doors and windows in particular—were painted onto the surface.

William Henry Bartlett's watercolour of the Prescott Gate was completed about 40 years after the Duberger and By model was shipped to Woolwich. Together, the two works hint at the construction the area underwent in the interval.

William Henry Bartlett, *Prescott Gate, Quebec*, watercolour, pencil and black crayon on wove paper, ca. 1845, e000996278

La vieille porte Prescott © Parks Canada, Québec, courtesy of Parks Canada

his scheme to justify to authorities in Britain the validity of certain elements of his defence plan. Apparently the colonial administrators had proceeded with some of these elements knowing they would never be able to defend the fortress from an American assault without them. Unfortunately, the committee of engineers saw things differently and had rejected some of the proposals. Mann likely used the model to justify the need for enhancements to Québec's defences and to convince the proper authorities why some changes were initiated without the prior approval of the commitee.

After the model arrived at the Royal Military Repository, an annex of the Artillery School at Woolwich, By confessed to the Inspector General of Fortifications, "I was as little sparing of my purse, as of my time, when off duty, to make the model not unworthy the Corps to which I have the honour of belonging." Why did he and Duberger make such a sacrifice? What were their motives? Possibly they saw this "princely toy" not as a chance to communicate needed improvements to Québec's defences, but as an opportunity to enhance their own careers in His Majesty's service. Certainly both Duberger and By requested such recognition when they submitted their applications for promotion soon after the model's completion—and neither man was disappointed. Duberger applied for promotion within months of the model's departure for Britain, and by 1813 he was commissioned an officer first class. Before the model's eighteen crates were packed and bound for London's wharves on board the brig *William*, By was promoted to captain.

But it was John By who seems to have received most of the recognition for their work. This has led to some speculation that he may have misrepresented himself and taken full credit for the project once he was safely back in England, well beyond the reach of anyone who could claim otherwise. In 1812 he was placed in charge of the Royal Gunpowder Mills outside London, in 1821 he was promoted to lieutenant colonel, and in 1826 he returned to Canada to superintend the construction of a canal system linking the Ottawa River with Lake Ontario via the Rideau River. His long-time mentor, Gother Mann, now an inspector general of fortifications at the Board of Ordnance, personally selected him for the job. The Rideau Canal would eventually prove to be By's greatest achievement and the one with which his name would be most readily associated.

It is difficult to assess what role the Duberger and By model played once it reached the Royal Artillery School in Woolwich, but it is probably safe to assume it must have had some effect on the Inspector General of Fortifications. Not only were all the participants in the project amply rewarded, but most of Mann's suggestions were eventually implemented in one form or another, although it took about two decades for everything to be completely in place. Within four years of the model's completion, British-American relations had deteriorated to such an extent that on June 18, 1812, the United States Congress formally declared war on Britain.

A contemporary view of the salon in the Château St. Louis, Québec, where the Duberger and By model was put on public display prior to its transport to Woolwich. The three-dimensional model offered a realism that everyone could easily comprehend and gave residents an aerial perspective of the city long before such views were technically possible.

George Heriot, *Dance in the Château St. Louis*, watercolour over pencil on laid paper, 1801, C 000040

By turning Canada into a battleground, the American government effectively brought to fruition the worst fears of Québec's colonial administrators. In the end, however, Québec's fortifications proved far too formidable for an American attack. Even though the American army outnumbered its British counterpart, it stayed well away from the fortress, a decision that in itself speaks well for Gother Mann's improvements. Instead, the Americans concentrated their assaults on key British installations along the upper St. Lawrence River and what was then widely regarded as Britain's western frontier—the Niagara Peninsula, the Detroit-Amherstburg region, and the upper Great Lakes. Sitting secure and unmolested over the course of the two-year war, Québec was essentially free to become "the door of entry for that force the King's government might find it expedient to send." Just as its administrators had predicted, Québec served the British well by acting as a centre for the distribution of munitions and military personnel to those parts of the colony bearing the brunt of the American invasion. ❧

MAP
of the City of
HAMILTON
in the County of Wentworth

CITY OF HAMILTON INCORPORATED VICTORIA 1847

Canada West
Surveyed and drawn
by
MARCUS SMITH
1850-1

DUNDURN.
The Residence of SIR A.N. MAC NAB.

GREAT WESTERN RAILWAY

BURLINGTON

COURT HOUSE

CHRIST CHURCH

Front of the Fourth Concession Township of Barton and South Limit of Corporation

CHAPTER 5

A Bird's-Eye Perspective

"THE BUSINESS HAD BEEN practically without competition as so few could give it the patience, care and skill essential to success," lamented bird's-eye artist Oakley Bailey to an American reporter in the 1930s. Then in his senior years, Bailey was reminiscing about the 1870s and 1880s, when he and a handful of itinerant artists wandered Canada's dusty roads selling inexpensive coloured lithographs of its cities and towns. "But now the airplane cameras are covering the territory,"

continued Bailey, "and can put more towns on paper in a day than was possible in months by hand work formerly." Almost without exception, the city views offered by Bailey and his colleagues were drawn as if the viewer were perched high in the clouds. From such a lofty elevation, the artists could show the street pattern, individual buildings, and major landscape features. They offered local residents a bird's-eye perspective of their community long before flight was a physical reality.

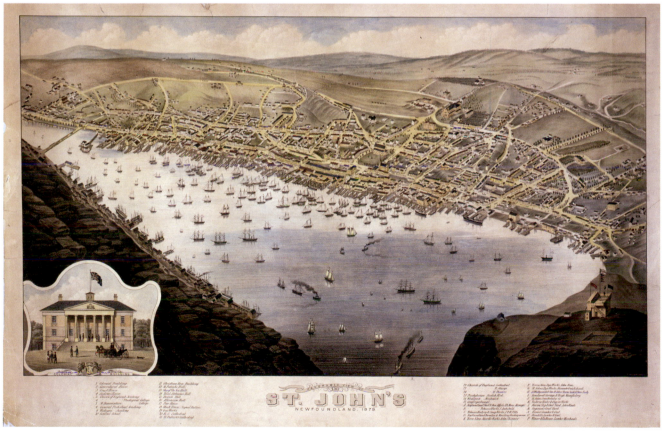

Since St. John's was competing with other east coast communities for the Atlantic shipping trade, it is not surprising that Albert Ruger's bird's-eye view of the city should focus on the busy harbour.

Albert Ruger, *Panoramic view of St. John's, Newfoundland*, 1879, C 006557

Canada in the late nineteenth century was well on its way to becoming a nation of city dwellers. Rapid population growth and the shift from a rural culture were slowly giving way to a new self-awareness that left civic leaders zealously competing with neighbouring communities for economic advantage and prestige. There was little cooperation as townsmen aggressively promoted their community in an attempt to translate opportunity into prosperity. City boosters searched for any gimmick that would attract trade and commerce at the expense of regional rivals, and coloured bird's-eye views were high on their list of "must have" promotional tools. Almost every community of any significance became the subject of a bird's-eye view, whether a small hamlet of several hundred like Simcoe, Ontario, or a city the size of Toronto. Nearly three hundred views of Canadian cities are known to have been produced over the seventy-year period ending around 1910.

Although the perspective was from a great height, the bird's-eye was still close enough to show some detail, but not so close as to reveal the city's imperfections. The result was a flattering portrait of late-nineteenth-century urban life: the sun is invariably shining, the season is eternally summer, and there is always a pleasant breeze to unfurl flags and send whiffs of smoke from factory chimneys. Everywhere there is order and everywhere there are signs of the city's vitality: factories running at full capacity, railway yards busy with speeding trains, streets bustling with horse-drawn wagons, sports grounds crowded with players and fans, and harbours clogged with shipping. Everything about the bird's-eye view screams success.

The ability of the bird's-eye view to embrace a city's finer qualities did not go unnoticed by local newspapers. "No better means of advertising our city or showing the good folk 'at home' the sort of place we live in could be devised," boasted the editor of a Halifax paper on seeing Albert Ruger's 1878 bird's-eye of the city. A decade later a front-page story in the *Victoria Colonist* declared that Eli Glover's view of the city "should adorn the walls of every office, store, and residence in Victoria." The newspaper found it "invaluable for the purpose of conveying to parties at a distance an accurate idea of the Queen City of the West." Given such

The bird's-eye perspective worked best for small towns like Berthier, Quebec, which may explain why so many of them were produced. The individual buildings, for example, are more clearly represented than in a similar plan of a large city.

Unknown artist, *Vue à vol d'oiseau de la ville de Berthier, P.Q.*, 1881, NMC 1412

sentiments, it is not surprising that bird's-eye views were among the single most popular form of mass-printed maps to come out of the nineteenth century, surpassing even the county maps (chapter 10).

Unlike most conventional drawings whose vanishing point is on the horizon, the bird's-eye view often situated its point somewhere in the sky. Although the result gave the mistaken impression that the city was located on a tilted plane, it at least allowed the artist to depict distant buildings almost as large as foreground structures. Since people were more likely to purchase a bird's-eye view if they could clearly make out their place of residence or business, the distortion added to the project's commercial viability, at least as far as the artists and publishers were concerned.

The artist began a bird's-eye view by moving slowly through a community making dozens of ground-level sketches of individual buildings and landscape features. Edwin Whitefield's sketchbook for Hamilton, which is the only known surviving example, shows us that the artist had already figured out from what angle he would view the city before he began sketching. For example, his building sketches were made as they would be seen in the final drawing. In other words, buildings on the west side of the city (which became the left side of the print) were sketched from the southeast, and buildings on the east side (the right side of the print) were sketched from the southwest. Since the viewer of the lithograph would be looking north, buildings on the north side of the street received more attention than those on the south.

Whitefield probably used Marcus Smith's map of the city, which had been published just two years earlier in 1850. With this map as a base, he would have been able to construct his perspective grid, taking into account the orientation and elevation he had determined. Using his field sketches, he would redraw every street on the perspective, taking care to make each building the correct size and to locate it properly within the block. From his preliminary drawing he would then prepare a more attractive and carefully coloured final drawing, which would be displayed around the city in an effort to attract advance sales for the lithograph. This final drawing would have been sent to the lithographer, Endicott and Company of New York, to be copied onto a stone for printing. As

A view of James Street, Hamilton, in the 1860s, showing the main part of the city as it would have appeared when Whitefield wandered the city's streets sketching its buildings.
Unknown photographer, *James Street, Hamilton*, ca. 1860s, C 008860

Whitefield did the preliminary sketches for his Hamilton bird's-eye in a small notebook. His sketch for James Street shows the buildings from the perspective that would be used in the finally drawing.
Edwin Whitefield, *James Street*, pencilled sketch on paper, 1854, C 036230

Marcus Smith's map of Hamilton was published in 1850, just prior to Whitefield's visit, and was probably used by him when preparing his bird's-eye view of the city.

Marcus Smith, *Map of the city of Hamilton in the county of Wentworth, Canada West,* Mayer and Kroft, 1850, NMC 11371

Oakley Bailey pointed out, the work required a great deal of patience and care. It had to be accurate, or the local citizenry would criticize the work, but not so precise as to show the city's imperfections or townsmen would be unlikely to buy it.

Publishers of city views almost always attempted to limit their financial risk by securing advance subscriptions from citizens, local businessmen, or other purchasers. Albert Ruger's 1878 print of Charlottetown, for example, was said by the local newspaper to cost "the sum of $500 … in order to bring" the lithograph "to its perfect condition." Ruger had to sell only two or three hundred advance subscriptions (the usual print run) at $2 or $3 each to pay for the printing and earn himself a nice profit.

There were other ways by which artists and publishers (they were not always one and the same) could regain some of their costs. Some publishers included marginal notes with the names of businesses, industries, professional offices, and prominent citizens, all of them no doubt included on payment of an extra fee. Other bird's-eyes added the name of a business or industry across the roof or building façade. Still others filled the margins of the print with vignettes of private residences, businesses, and public buildings. The *Advance* challenged community leaders to purchase

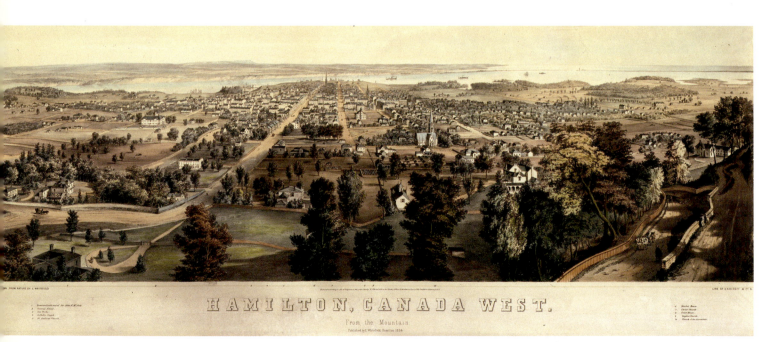

Whitefield's bird's-eye of Hamilton was published in 1854. Later bird's-eye artists generally used a higher perspective for their city views, which allowed them to include a greater proportion of distant buildings.

Edwin Whitefield, *Hamilton, Canada West, from the mountain,* New York, Endicott and Co., 1854, C 010662

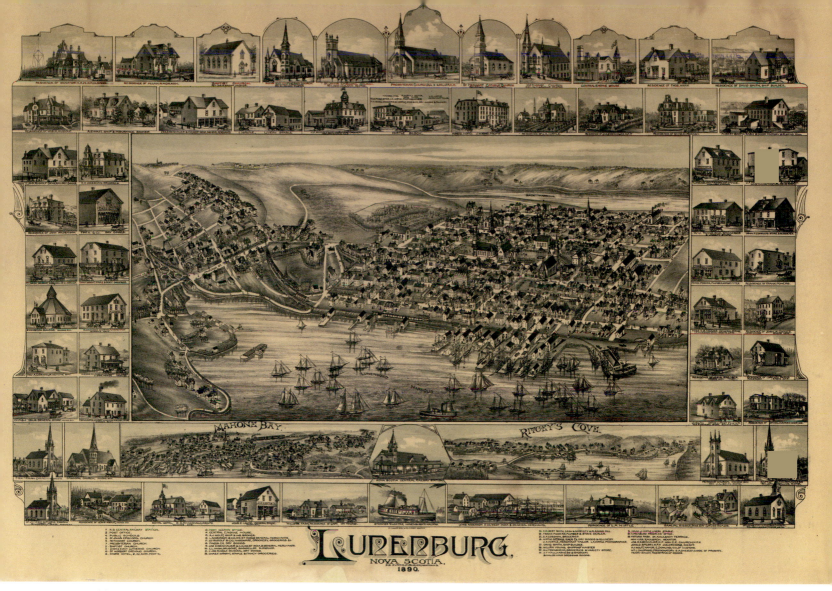

Public support for the bird's-eye industry is sometimes evident in the number of sketches that border a map. More than 60 vignettes frame this bird's-eye view of Lunenburg, Nova Scotia. The added revenue these views brought the publisher, D.D. Currie of Moncton, helped to subsidize production costs.

Unknown artist, *Lunenburg, Nova Scotia*, Moncton, D.D. Currie, 1890, NMC 18414

one of the vignettes that A.M. Hubly was offering in his bird's-eye of Chatham, New Brunswick, "for advertising purposes as well as characterizing the businessmen of the place." It went on to call the vignettes "a commendable feature which should be encouraged."

City councils occasionally helped subsidize some of the costs of publication. Although asked to purchase a thousand copies for distribution across North America, the city council in Port Arthur decided instead to support the venture by underwriting the cost of a vignette of the new school. Winnipeg's city council, on the other hand, helped Thomas Fowler to underwrite some of his costs by purchasing one hundred copies of his 1880 bird's-eye of the city. Apparently these were sent to "a number of gentlemen in England, Ireland, and Scotland, who will be able to use the views to excellent advantage in promoting the interests of

the city." A number of copies were also mounted on rollers "for the purpose of having them hung up in the cabins of ocean steamers, and in the sitting rooms of some of the principal hotels in the old country." Likewise, the Dawson City council is on record as having purchased one thousand copies of H. Epting's 1903 bird's-eye, which it mailed to "the high officials of the Dominion, the members of Parliament, the mayors and leading citizens of all the larger cities of Canada and the United States, and to many of the great newspapers of the country."

Unfortunately, we have no idea how successful the distribution of views may have been in stimulating urban growth and commercial development. As urban geographer John Reps has pointed out in his monumental study of North American bird's-eye views, their effectiveness in accomplishing the goals of community

COLOUR LITHOGRAPHY

The bird's-eye industry probably would not have been possible without the invention of lithographic printing. The process called for the sketch to be transferred onto the surface of a piece of polished Bavarian limestone using a crayon made from a combination of wax, soap, tallow, and shellac. When the transfer was completed, the stone was dipped in a bath of diluted nitric acid and water so as to fix the crayon and keep it from spreading. The stone was placed in a press and rolled with ink, which easily adhered to the crayon but was repelled by the wet stone. Paper was then placed on the stone and pressed against it with a heavy roller so as to transfer an imprint of the image. The process was repeated for each print.

Lithography was considerably cheaper than copperplate engraving (chapter 7), and by the 1860s it was predominant. The result was the same as in copperplate printing, but the technique differed: engraving was mechanical; lithography, chemical. It took less time to transfer the drawing to the stone, repairs and corrections to the crayoned image could be undertaken without completely redoing the drawing, and the stone could be used over and over again.

Although the earliest bird's-eye views were hand-coloured, by the 1860s it was possible to run coloured inks through the press and to achieve tonal effects. The process called for a different stone for each colour (sometimes as many as thirty stones might be used on a single print) and resulted in a more uniform colouring from print to print at a cheaper price.

Lithographs have been called "democratic art." For many 19th-century middle-class Canadians, their first artistic piece of value was a lithograph. It was not uncommon for auction houses of the period to advertise the lithographs that were to be sold at estate auctions.

Unknown artist, *Niagara Falls Park and River Railway*, Toronto, Toronto Lithographing Co., 1900, NMC 43318

Lithography had other advantages over copperplate printing that reduced labour costs. A typical stone was good for two to four thousand impressions before a new transfer had to be made, easily two to four times the number of pulls obtained from copper before an engraver had to redefine the incised lines in the plate. As well, the method of putting an image on the stone, although initially done by hand, was eventually automated through a photographic transfer process. Lighter, more durable zinc sheets came to replace the heavy lithographic stones, allowing the presses to be converted from flat-beds to high-speed rotary systems (chapter 9).

Prices for lithographed bird's-eye views varied from $1 for a view of Dawson City, sold through the town council, to $2 for a view of Paris, Ontario, to $15 for this 1893 coloured view of Brantford, Ontario.

Unknown artist, *City of Brantford, Can.*, Toronto, Toronto Lithographing Co., 1893, NMC 15029

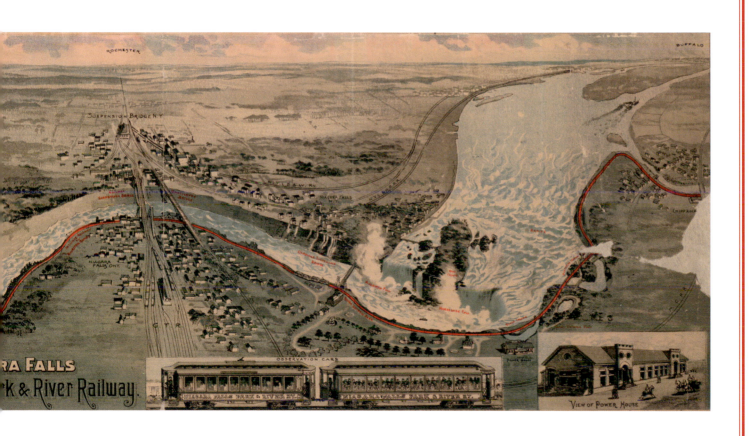

Birdseye View of Dawson, Yukon Ter., 1903.

Middle-class Canadians in the late 19th century hungered for lithographs to hang on their walls. They bought pictures of their political leaders, naval engagements, natural disasters, public buildings, wildlife, and maps—almost anything that could be presented as an image. But bird's-eye lithographs, like this 1903 view of Dawson City, were the single most popular category of mass-produced images.

H. Epting, *Birdseye [sic] view of Dawson, Yukon Ter., 1903*, Vancouver, B.C. Print and Engr. Corp., 1903, NMC 21044

boosters was probably overstated by sales agents and publishers. Surely "experienced business leaders and public officials must have realized that urban growth depended on more than circulation of a handsome civic portrait," concluded Reps.

Universally, newspaper editors across Canada gave the bird's-eye views favourable coverage. One noted that Albert Ruger "with un-wearied patience … has drawn every building in outline on a scale small enough to show the whole city; has located every building so exactly that every citizen can pick out his own place. Streets, railways, streams, public and private buildings seem to be quite correct." A Halifax newspaper expressed similar sentiments: "The picture …

presents clearly and distinctly every part of the city. Not only the streets but each house can be distinctly pointed out; while the public buildings can be distinguished in a moment … The view of the harbor too, is excellent, as much care being taken to shew each wharf as to outline the public and other buildings."

Similarly, the *Winnipeg Daily Times* described Thomas Fowler's view as a "marvellous work of art." A later article in the same paper declared the work "useful as well as ornamental." "It is beautifully finished," continued the editor, "the artist having evidently taken great pains in producing a handsome and faithful representation of the city—a view that is much better than an ordinary map or chart, inasmuch as every building is plainly shown and is easily distinguished." The *Victoria Colonist* had equal praise for the careful attention to detail shown by Eli Glover: "The view [of Victoria] is … supposed to be taken from a considerable elevation and thus every street, public building, and private residence is depicted with extraordinary accuracy. The picture … contains almost photographic representations of more than 2,500 buildings."

This last comparison to a near "photographic representation" is common with newspaper editors, so much so that one can not help wondering why the artists troubled themselves with drawing and did not use photography instead. Throughout the late nineteenth century photographers did make an attempt to capture Canada's cities and towns on film but lithography had some distinct advantages. Where photographers were limited to finding a real vantage point (not always possible in some towns because of jutting buildings, steeples, hills, or other obstructions), bird's-eye artists could always find an imaginary viewpoint that was higher and more distant, enabling them to capture the entire city within that view. More important, the bird's-eye artist could offer a coloured image and take liberties that no photographer could ever hope to emulate.

When bird's-eye artist Edwin Whitefield was asked why he did not use photography in his compilations, he admitted that it was not really an option because it limited his ability to play with the view: "Very seldom would this [photographic] process have been desirable," he wrote in 1886, "because in very many cases, alterations and additions have been made at various times to these buildings, which I have left out, and shown them as they were originally. Then again, frequently trees hid important features; of course in sketching I could leave these trees out partially or wholly." He also thought nothing of leaving out buildings that were less attractive. In his Hamilton sketchbook, for example, a small notation on the page used to record the homes along Maiden Street indicates that the end of the street contained "no homes, little shanties." Interestingly, there is no hint of the shanties in Whitefield's final lithograph.

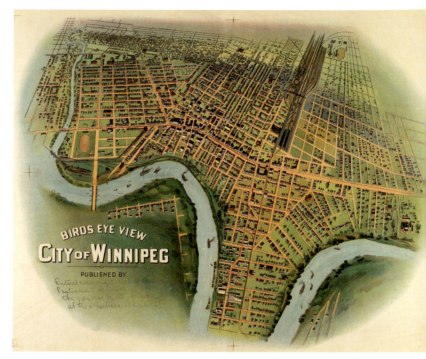

Colour lithography was introduced to Canada just when the country was shifting from a rural-based economy to an urban one. Vibrant bird's-eye views, like this one of Winnipeg in 1900, helped civic leaders present their communities as great places to live and do business.

Bulman Bros., *Birds eye [sic] view, city of Winnipeg*, Winnipeg, Bulman Bros., 1900, NMC 23821

Eventually photography did lead to the demise of the bird's-eye lithograph. The panoramic camera of the early 1890s and the aerial camera of the 1920s could capture in a single day what took Edwin Whitefield, Oakley Bailey, and their fellow artists months of footwork. The airplane in particular offered an advantage of which the bird's-eye artist had only dreamed.

Although the bird's-eye lithograph is a remarkable testament to Canada's early urban development, it is perhaps wise to remember the words of the *Yukon Sun* in describing the Dawson City bird's-eye view as "one that attracts more than passing attention, and answers perfectly the purpose for which it was designed—the advertising of Dawson." In other words, we must be willing to accept these relics for what they are: not an account of how things were, but a version of how the artists and their sponsors wanted them to appear. ꙮ

CHAPTER 6

Mapping for Fire

"THE BUSINESS OF FIRE INSURANCE has become so systematized in this country that correct maps have been made by competent engineers," wrote Henry S. Tiffany in his 1887 instruction manual for fire insurance agents. Written "from a practical standpoint," Tiffany wanted to encourage "those just commencing in the insurance business" to embrace the special maps that had been prepared for agents working in the field. The maps had been available in Canada for more than a decade and, according to Tiffany, were "a very great convenience in the prosecution of the business, and their use simplifies and lessens the work very materially." If agents chose to ignore the published maps they had their work cut out for them, because insurance companies then expected agents to do their own surveys and pre-pare scaled diagrams that would help underwriters to calculate fire risk. As far as Tiffany was concerned, it was a waste of an agent's time and energy to

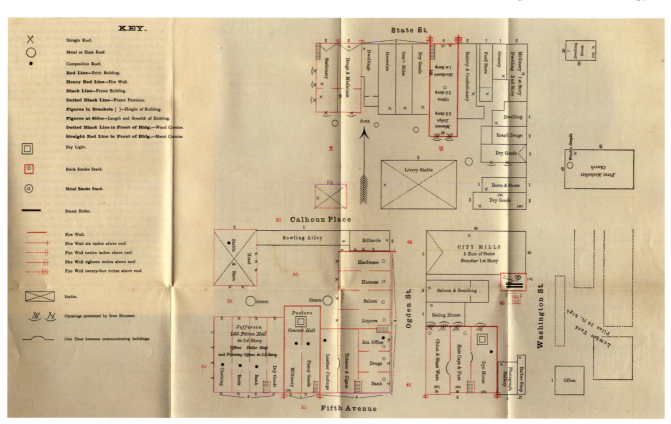

Tiffany encouraged agents to use the specially prepared fire insurance maps that were published by Goad and others. If such maps were unavailable, agents were expected to provide underwriters with detailed plans they had prepared themselves, similar to the one shown here.

Henry S. Tiffany, *Insurance plan*, from *Tiffany's Instruction Book for Fire Insurance Agents Together with a Digest of Depreciations and Rules for Ascertaining the Cost of buildings by their Cubic Contents*, 25th ed., Chicago, H.S. Tiffany and Co., 1887, p. 80

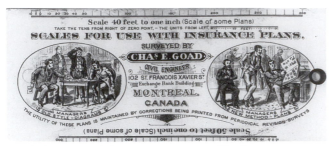

Goad's business circular from 1888 promoted his fire insurance plans by stressing their ease of use for insurance agents and adjusters alike. Goad's plans provided insurance underwriters with the uniformity and clarity they required for their fiscal survival.

Charles E. Goad, *Scales for the use with insurance plans,* Montréal, Chas. E. Goad, 1888, C 061913

In the late nineteenth century, fire was Canada's urban plague. Although few lives were lost, many were certainly ruined. The older eastern port cities with their narrow streets and densely packed warehouses and factories were particularly hard hit. It is estimated that before the Great War, eastern Canada suffered through more than three hundred major fires and many cities were hit more than once: Montréal experienced five major fires, Saint John, seven, and Québec, thirteen. By 1918 the direct costs of these losses affected Canadians to the extent of $700 million, while indirect costs in the form of interrupted business and loss of employment earnings and tax revenues were "beyond the power of figures adequately to represent."

Since two-thirds to three-quarters of the population lived and worked in wood-frame buildings, with heating and lighting relying on open flames (candles, oil lamps, coal furnaces, and wood stoves), Canadian cities had all the ingredients to turn almost any flare-up into a calamity. In the event of a major fire,

undertake such mapping when "perfectly correct" maps were already available. As well, since agents had no special training in mapmaking, there was always a chance they might make an error that could cost their company dearly. Mapmaking for the insurance industry was a serious business because fire was serious business.

Jules-Ernest Livernois, *Fire in lower town, Quebec,* 1866, C 004733

Before the First World War, Canada claimed the dubious reputation of having the world's highest per capita loss from fire. The blazes left as many as 20,000 people homeless at a time, created enormous relief problems, and tested social programs to the limit.

Unknown photographer, *View of the ruins of the Anglican cathedral of St. John the Baptist, Newfoundland,* 1892, PA 066621

William and Thomas Wentworth's lithograph of the fire that swept through Saint John's commercial district and in a single night reduced a five-block section to a pile of smouldering ash. Forty years later, the New Brunswick capital was hit by a more devastating fire that left 13,000 people homeless and bankrupted several insurance companies.

William H. Wentworth and Thomas H. Wentworth, *View of the great conflagration that took place on the night of Saturday, 14th of January 1837 (St. John, New Brunswick)*, lithograph, Boston, Thomas Moore's, 1838, e000943165

policyholders across the country helped pay for the rebuilding. Grove Smith, in his 1918 study, *Fire Waste in Canada*, commented: "It is this comity of interests through fire insurance that enables recovery from the effect of fire … A loaf of bread bears the cost of insurance upon the buildings and stock of a retail store, bakery, flour warehouse, flour mill, terminal elevator, country elevator and farmer's barn."

The system worked only if insurance companies kept their exposure to fire risk evenly spread across a city. When fire destroyed a major portion of Saint John in 1877, the $29 million paid out by insurance companies bankrupted six of them given the high concentration of their policyholders in the wasted area. What these firms needed was a simple way of knowing the physical characteristics of the buildings they were expected to insure and the number of their policyholders.

Surveyor D. A. Sanborn of Somerville, Massachusetts, was the first to recognize the importance of maps in meeting the special needs of Canada's underwriting industry. A decade after establishing his insurance mapping business in New York City in the early 1860s, his crews ventured across the border to survey fifteen of Canada's eastern cities: Belleville, Brockville, Brantford, Cobourg, Guelph, Hamilton, Kingston, Lindsay, London, Ottawa, Peterborough, Port Hope, Québec, St. Catharines, and Toronto.

One of his locally hired surveyors was an engineer by the name of Charles E. Goad. Born in London, England, in 1848, Goad claimed to have received an "arts" education at the University of Oxford, although there is no record of his graduating. Almost immediately on immigrating to Canada at the age of twenty-one, Goad "engaged" in railway construction, either in the capacity of a draftsman or engineer, first on the Toronto, Grey and Bruce Railway (1869–73), then on the Montreal Northern Colonial Railway (1873–75), and finally on the Halifax and Cape Breton Railway (1877–78). Apparently with Sanborn's blessing and while still employed in the railway business, Goad undertook a

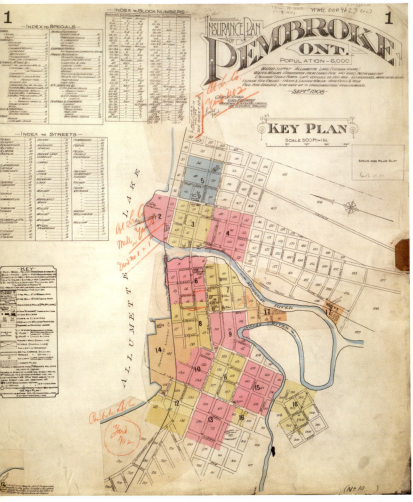

Goad's title page and legend from his 1908 insurance plan of Pembroke, Ontario, shows how he organized the extensive information he collected on each community as part of his insurance mapping.

Charles E. Goad, *Insurance plan of Pembroke, Ont.*, Montréal, Charles E. Goad, 1908, p. 1, key plan, NMC 9623

printed on sheets measuring 53 by 63.5 centimetres and at a scale of 1:600 (1 cm to 6 m)—just the right size for agents to write in the policy number of the buildings they insured so as to have a visual record of the concentration of policyholders. Coverage for small communities might be issued on just one loose sheet, while that for bigger cities, such as Toronto or Montréal, might run to nine volumes. (In the 1950s, large urban areas required as many as twenty volumes.)

Goad's maps evolved into documents that displayed a wealth of information about the physical characteristics of individual buildings, its fire-fighting equipment (hydrants, hoses, and extinguishers), and its contents (type of furnaces, fuel, and lighting), especially if it was at higher risk for fire. Through a combined use of colour coding, symbols, and abbreviations, the Goad plans show a building's wall construction (red for brick, yellow for wood frame, blue for stone or concrete, and buff for buildings constructed mainly of fire-resistant materials) and roof composition (slate, tin or metal, wooden or mortar shingles, and skylights), while a set of names and abbreviations record the building's use ("D" for dwelling, "S" for store, "Ten's" for tenements, and so on).

Since the surveys were so detailed and the production runs small (the number of insurance companies operating in Canada during Goad's lifetime was never more than thirty), the maps were relatively expensive. In 1880 a single-page plan of a small Ontario town sold for about $5, while all fifteen sheets that comprised a larger urban centre sold for $55. Given that a good quality city map at the time might have cost 10 to 50 cents, Goad continually had to justify his prices to underwriters. In an effort to keep prices down and to ensure that individual plan holders kept their copies up-to-date, Goad devised a rather unique marketing system. Rather than being sold to users, the plans were issued on a subscription basis with all subscribers sharing in the costs of production. Subscribers were also promised a share of any profit from future distributions of a plan once expenses were met. According to Goad's corporate records, however, a redistribution of profits seldom materialized.

By retaining ownership, Goad could recall outdated plans from insurance offices when new editions became available. (Fortunately, some companies

survey of Lévis, Quebec, a city that Sanborn had neglected and one that had been ravaged by fire several times over. The Lévis survey was published in 1876, and since there appears to have been sufficient encouragement from insurance underwriters, Goad quit his railway job in 1878. He had already bought out Sanborn's Canadian business and established a small office for himself in downtown Montréal the previous year.

The style Goad adopted for his fire insurance plans followed the standard of the day and borrowed heavily from the Sanborn company, which in turn copied the plans first published by William Perris, an English engineer who produced the first detailed fire insurance map of a North American city, New York, in 1852. Goad's plans were hand-coloured lithographs, normally

ignored Goad's instructions, with the result that today's researchers have access to an invaluable archival resource.) At first Goad distributed his plans in sets of two: one for a company to retain in its head office and a second to send to its local agent. Local agents had only to refer to a plan number and building address in their reports to head office. Underwriters could use the information on the plan to assess the risks associated with insuring a particular structure.

The system was a constant source of frustration for Goad. As many agents, particularly in smaller towns, represented more than one company, they received several copies of expensive plans, some of which were passed along to other agents of companies who had contributed nothing towards their production. Some local agents were also in the habit of lending plans to colleagues who made whatever tracings they required, again without compensating Goad. Once he realized what the local agents were up to, Goad realigned his prices. He began charging higher fees for each head office copy but offered to send out the local agent's copy free. By communicating directly with local agents himself, Goad could ensure that they did not receive multiple sets of his plans. However, there was not much he could do about agents who were willing to share their plans for tracings, other than to plead his case and call upon their sense of fairness.

Goad recognized that his plans had to be kept up-to-date or they ran the risk of becoming obsolete. "Changes and improvements are being made daily in the construction of villages, towns and cities ... and in the means of protection against fire," admitted Goad in his monthly journal, *Insurance Society*. "Insurance men of all classes must ... keep themselves posted in these matters, and as personal observation is impossible, the facts and conditions of such changes must be gathered for them." Goad placed each plan on a two- to six-year cycle; rather than reissue a completely new map, each time changes were required he simply printed them all together as patches. These "patch pages" were sent to holders of the original plans, who cut out the patches and glued them onto the appropriate plan. The revisions were sometimes extensive. In one instance he issued revisions to sixteen towns that amounted to 512 patches and 593 pen corrections. A completely new edition of a plan was issued only when the revision slips began to interfere with readability.

Goad's devotion to the fire insurance industry was unparalleled for his time. In 1881 he moved his offices to a much larger facility on St. François Xavier Street in Montréal so that he could provide underwriters with a general meeting place, and more important, a reading room that contained "standard insurance literature." The same year, in an effort "to assist in the introduction of a more rational, more systematic, and more profitable means of conducting business," he also started the monthly journal *Insurance Society*. Through a "mutual interchange of information," Goad hoped that the publication would lead not only to increased cordiality in the highly competitive industry but also to an "improvement in fire protection in cities, towns and villages."

His dedication to the industry paid off financially. By 1888 Goad could boast that his plans extended "from the Atlantic to the Pacific, embracing every plan that has been asked for by two or more Companies; and in this the fifteenth year of continuous work, comprises 423 surveys of compactly built business centres and surroundings, about 56,000 acres in area, or 88 square miles." More significant, his plans were a boon to insurance profits. Since his plans had been introduced, companies that used them systematically were reporting an average 8 percent profit, those that used them "spasmodically" realized no profit, and those that made no use of them experienced a 43 percent loss. Euphoric about his success in Canada, Goad opened an office in London in 1885, which soon became a base for expansion into Europe, southern Africa, the Caribbean, and South America.

Following Goad's death in Toronto in 1910, his sons signed an agreement with the Canadian Underwriters' Association that allowed the Charles E. Goad Company to produce and revise fire insurance plans for the Association. Goad's fifty-year domination over insurance mapping in Canada came to an end on October 26, 1917, when the Underwriters' Survey Bureau Limited was created and took over all map production. At the time of the transfer, Goad had mapped some 1,300 Canadian cities, towns, and villages. In 1931 the firm ceded to the Bureau all rights to its Canadian plans in order to concentrate on its London-based operations. Although the London office of the Charles E. Goad

GOAD MOVES TO LONDON

When Charles E. Goad established the London branch of his fire insurance business in 1885, he was the first Canadian cartographer to take his mapmaking expertise overseas. The venture was not without its challenges. In 1888, for example, the *South Wales Daily* reported on two young gentlemen who were reportedly working for a "combined insurance company." It seems they were calling on tradespeople with a request to survey and sketch their back premises and take stock of the surrounding walls. The editor thought that "this plan has been adopted to further the ends of contemplated burglings" and advised his readers to be wary of the pranksters. Goad's representatives were quick to respond with an explanation about the nature of their business. Although the editor eventually retracted his statement, the damage was done. Other newspapers soon repeated the warning and, as a result, it was some years until Goad's surveyors had unrestricted access to buildings for a fire risk assessment.

Goad's perseverance paid off and he was soon reporting a monopoly on fire insurance mapping in the United Kingdom as well. His success did not go unchallenged, however. In one instance some of

In addition to his fire insurance plans, Goad also maintained a meeting room in his Montréal headquarters for "the fraternity of Underwriters." Furnished with a library of standard insurance literature and leading journals, the room was to act as a central point for the collection and diffusion of information. "My motive is not pecuniary profit," announced Goad to his customers, "other than that which may accrue to me ... by inducing a healthier and more prosperous state of affairs in underwriting circles."

Unknown photographer, *Charles E. Goad*, ca. 1880, C 076036

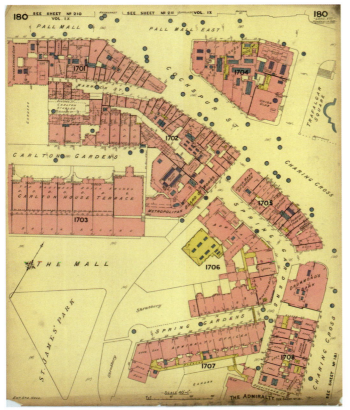

Goad's carefully prepared plan of Trafalgar Square helped establish his reputation as England's leading insurance mapmaker. "Were it not for the admirable series of fire maps," concluded the British Fire Prevention Committee in 1902, "the public would certainly have to pay higher premiums than is at present the case."

Charles E. Goad, *Fire Insurance Plan of London*, London, Charles E. Goad, 1888, p. 180, NMC 30086

Goad's former employees began offering the same service, but at less than half the rate. While letting on that he favoured competition, Goad argued that there should be only one insurance mapping company—his—because it was already difficult enough for his representatives to gain access to English businesses. If two or three competitors were to knock on the same door, the problems of access would only increase. "One organization should suffice," stated one of Goad's trade circulars. "If two or three competitive plans are made, the difficulties of admission will be enormously increased that careful surveyors will not continue on the work, or may omit important points, because disheartened by repeated and persistent refusal to admit."

His argument seems to have worked. Despite the lure of cheaper prices from other firms, most British companies preferred to stay with Goad. Within two years he was advertising the availability of plans for fifteen cities: twelve in England, two in Scotland, and one in Ireland. He used the London office, which eventually became his headquarters, to expand overseas. By the time of his death in 1910, he could boast of having mapped cities in Chile, Denmark, Egypt, France, Mexico, Mozambique, South Africa, Turkey, and the Caribbean.

The title page from Goad's fire insurance atlas of Istanbul (Constantinople) demonstrates how far reaching his insurance mapping business had become.

Charles E. Goad, *Plan d'assurance de Constantinople*, London, Charles E. Goad, 1904, title page, vol. 1, NMC 14958

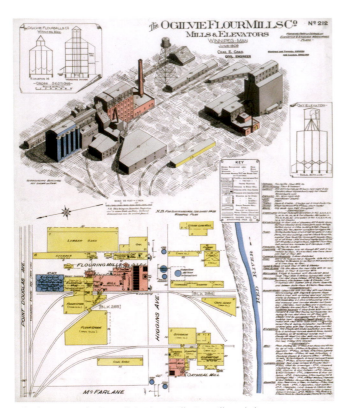

Charles E. Goad, *The Ogilvie Flour Mills Co. mills and elevators, Winnipeg-Man. June-1909*, Montréal, Charles E. Goad, 1909, NMC 10577

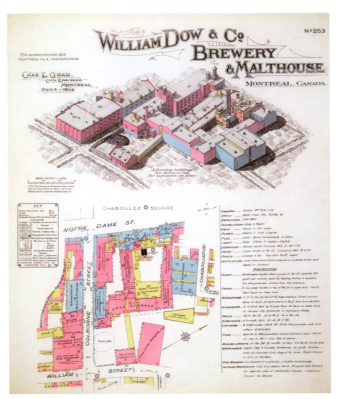

Charles E. Goad, *William Dow & Co. brewery and malthouse, Montréal, Canada*, Montréal, Chas. E. Goad, 1903, NMC 10582

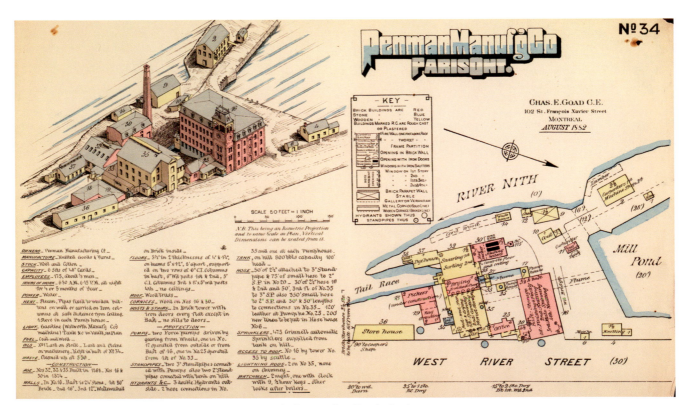

Goad often provided separate plans of industrial sites, which typically offered a bird's-eye perspective of the site as well as a vertical plan. In order to minimize their risk, a number of companies insured different parts of each complex.

Charles E. Goad, *Penman Manuf'g Co., Paris, Ont.*, Montréal, Chas. E. Goad, 1882, NMC 9618

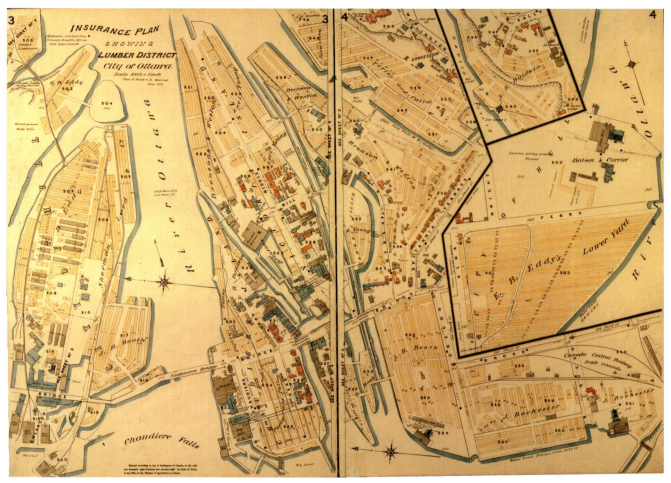

Fire insurance plans have been called the "footprint" of the Industrial Revolution. Like this plan of Ottawa's lumber district in 1878, they bear witness to Canada's industrial development from waterpower to steam to combustible engines. Nowhere can one find a better record of more than a century of urban growth.

Charles E. Goad, *Insurance plan showing lumber district, city of Ottawa*, Montréal, Chas. E. Goad, 1878. From *Insurance plan of Ottawa, Ontario, June 1878*, NMC 10731

Company is still in operation, its mapping activities focus primarily on shopping centres.

In Canada the Underwriters' Survey Bureau continued using the same system established by Goad until its dissolution in August 1967. The Bureau concentrated its operations in eastern Canada (the Maritimes, Ontario, and Quebec) and left western Canada to the Western Canada Underwriters' Association and the British Columbia Underwriters' Association. With the merger of the three associations in 1960, all mapping operations were brought under the Underwriters' Survey Bureau. Just seven years later, the Bureau was absorbed into the Canadian Fire Underwriters' Association, which was renamed the Insurers' Advisory Organization in 1975. That same year, 101 years after the publication of the first plans by Sanborn, all

fire insurance mapping in Canada was brought to a close in favour of more affordable computerized systems for the collection and tracking of liability information.

Fire continued to plague Canada's urban areas well into the 1930s and 1940s, when building codes finally began to improve, fire-resistant materials were developed, and fire-fighting equipment became more accessible. Thanks to Goad's plans, however, the insurance industry was better prepared to sustain itself in the event of a major conflagration. But Goad's greatest legacy was unintended. His detailed plans, which he so desperately tried to control, are now widely sought by all types of researchers interested in using their uniformity and clarity to document aspects of Canada's early urban fabric. ❧

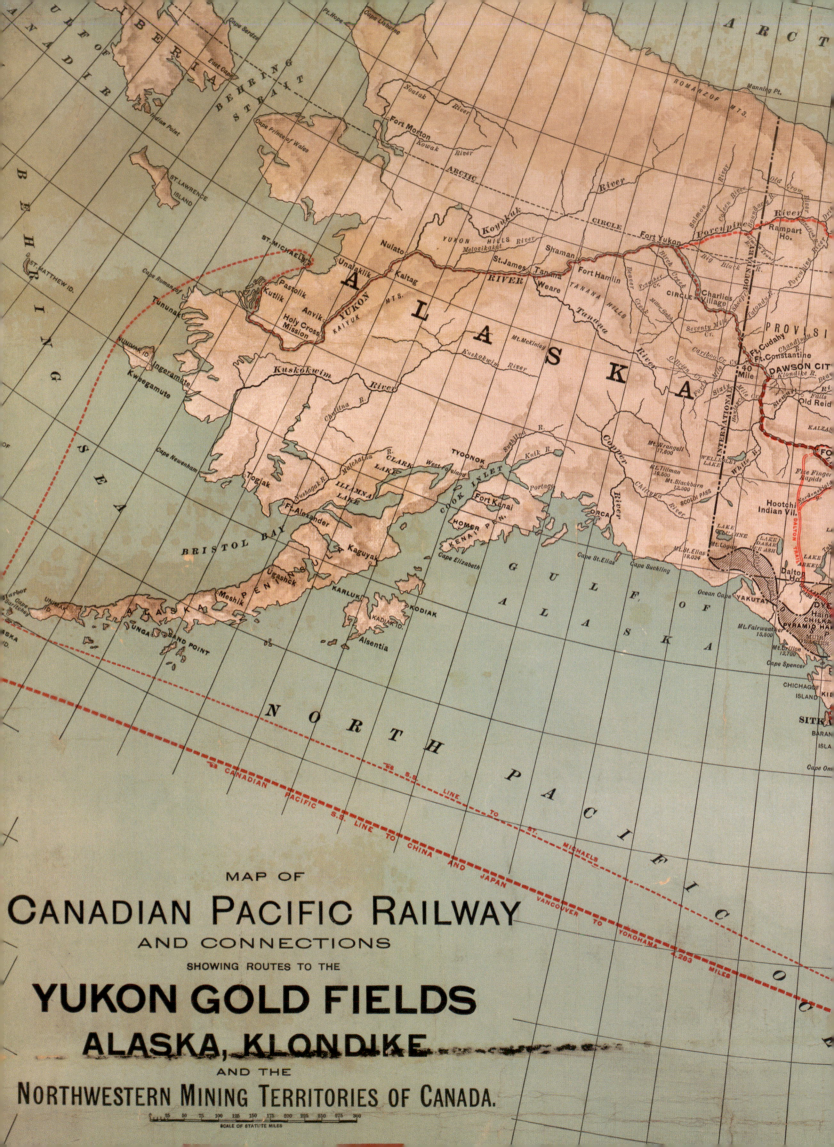

MAP OF
CANADIAN PACIFIC RAILWAY
AND CONNECTIONS
SHOWING ROUTES TO THE
YUKON GOLD FIELDS
ALASKA, KLONDIKE
AND THE
NORTHWESTERN MINING TERRITORIES OF CANADA.

SCALE OF STATUTE MILES

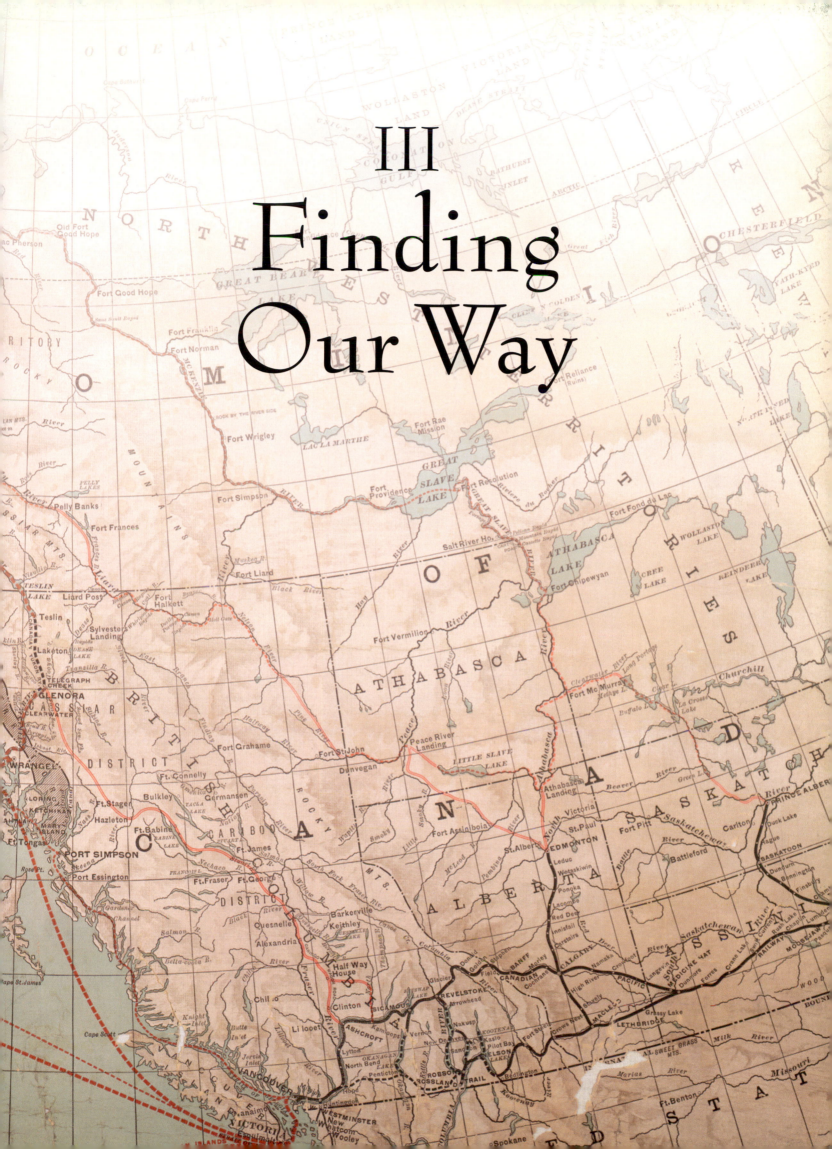

III
Finding
Our Way

Transportation by air, land, and sea has been crucial to Canada's development. Indeed, Canada may be the only country in the world whose very existence as a nation was contingent on building and maintaining a national railway—a provision that was intrinsic to the British North America Act. Where most countries look to revolutions and wars as defining moments in their development, Canadians take pride in a transcontinental rail line, a land-based substitute for the Northwest Passage that had eluded so many earlier travellers. It might be said that most of Canada's development has been a result of the need to conquer the divisiveness of space. This need has seen the introduction of increasingly faster and more reliable systems of transportation that facilitate the safe movement of people and goods.

At first, the transportation network developed slowly. With most Canadians living next to shorelines and rivers, before the nineteenth century people and freight moved by water. Land transportation was limited to a few macadamized roads that bypassed sections of water routes dangerous to navigate. With the development of canals and small feeder railways in the early nineteenth century, the network gradually extended away from the waterways and opened more remote interior regions to development. Communities mushroomed around these systems and defined their economic well-being by access to transport links. By the 1890s, hundreds of railway charters had been granted as industry attempted to complement navigation on the Great Lakes and the two coasts and extend the tentacles of commerce to Canada's hinterland, in particular the western Prairies. As automobiles became more reliable in the early 1900s, roads and highways were also added to what was already a complex mixture of transportation modes. Today, the airplane is one more means of conveyance that makes it possible for Canadians to live, work, and play in any part of the country, regardless of its remoteness.

Canada's mapmakers have always been quick to take advantage of each new method of transportation with products that both encouraged travellers to undertake difficult voyages and to assist them as they navigated their way across the country. This section looks at three

Gold rushes have a way of bringing out the best and worst in mapmakers. As would happen with the Klondike stampede half a century later, the Du Loup Gold Company issued map pamphlets showing the easiest routes to its placer mines in Quebec's Eastern Townships, thereby hoping to attract prospectors from Boston.

Du Loup Gold Company, *Du Loup Gold Company's Hotel Liniere Canada East*, ca. 1860, NMC 1059

Georges F. Baillairgé's survey of possible routes for a year-round road that would help people find their way from the St. Lawrence to Grand Baie on the Saguenay also provided the colonial government with an assessment of the region's economic potential.

Georges Frédéric Baillairgé (also Baillargé), *Chemins Baie St-Paul et Malbaie à Grande Baie*, 1862, NMC 1237

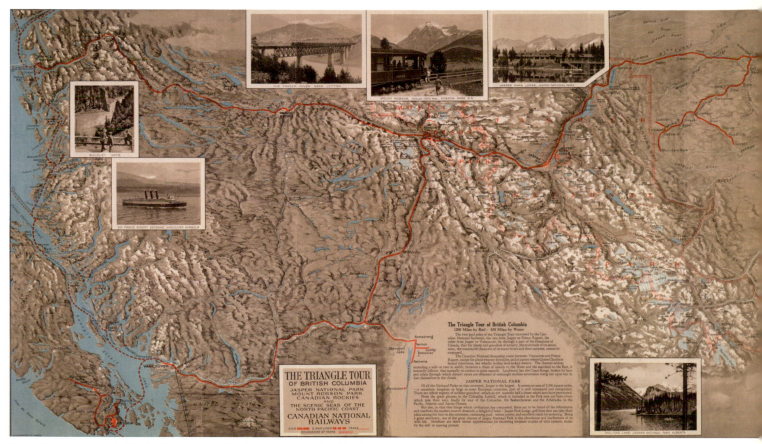

The railroads were probably the first to recognize the economic spin-offs of attracting tourists to Canada and unashamedly issued maps that flaunted Canada's attractions and the rail connections to them. This tour map of British Columbia by Canadian National Railways conveniently forgets to mention the existence of another rival rail company—Canadian Pacific Railway—that also provides service to Vancouver and the west coast.

Canadian National Railways, *The triangle tour of British Columbia, Jasper National Park, Mount Robson Park, Canadian Rockies, and the Scenic Seas of the North Pacific Coast,* Canadian National Railways, 1927, NMC 126592

types of maps that have helped Canadians to find their way. The first, created by a Swiss-born engineer in the employ of the British Army, was undertaken in an effort to improve safe navigation along the eastern seaboard. Launched shortly after the British conquest of New France, a preliminary set of charts took fourteen years to complete; the final set, another five. In the end, *The Atlantic Neptune*, as it became known, provided seafarers with their first accurate depiction of the North American coast from the Gulf of St. Lawrence to the Florida panhandle.

Maps are both an aid and an incentive to travel. The second chapter focuses on maps created by commercial publishers in the final years of the nineteenth century. Part of the promotional hype surrounding the Klondike gold rush, they were enormously influential in encouraging people to make the backbreaking trip

to the remote and unknown Yukon. Under the guise of providing accurate information to would-be prospectors, the maps proved to be deceptive, misleading, and highly optimistic.

The third chapter considers the maps created in response to the introduction of the automobile. Aided by technological advances in printing and paper-making, cartographers were able to create maps for the auto industry that were less and less expensive to produce and simpler to read. Provincial governments also helped this budding form of transportation by installing road signs that, with a map in hand, made it easier for drivers to navigate the network of provincial highways. By the 1950s, the free road map was an everyday feature of auto travel. Canadians had come to the realization that, with a good map beside them, there are no unknown roads. ☙

CHAPTER 7

Charting Eastern Waters

KNOWLEDGE OF THE ATLANTIC COAST—its shoals, reefs, currents, harbours, and anchorages—was paramount to the English-French struggle for North America. When mounting their amphibious assault on Québec, not even the French charts they had acquired after their capture of Louisbourg gave sufficient details to permit the British fleet safe passage up the St. Lawrence River. The lack of detailed charts was not a reflection of the inabilities of French surveyors:

the Dépôt des cartes et plans de la marine was well directed and its employees, who possessed surveying skills of the highest calibre, far outnumbered their British counterparts. Britain's entire Corps of Engineers in the eighteenth century, for example, never totalled more than seventy, of whom no more than seventeen were posted to North America—a pitifully small number compared to the corps of three hundred that served France.

Capt. Charles Ince's engraved view of the British siege of Louisbourg, where Joseph F.W. Des Barres and James Cook first proved their capabilities as surveyors by providing senior command with accurate plans of the French defences, including the well-protected harbour.

Charles Ince, *A view of Louisburg [sic] in North America, taken near the light house when that city was besieged in 1758,* hand-coloured engraving on laid paper, 1762, C 005907

Many French were as dismayed by the quality of available charts as everyone else. In one communication, the Marquis de Chabert complained that some French charts differed by as much as nine degrees of longitude for the same position. In the region of the St. Lawrence, this error amounted to a difference of some 600 kilometres—more than enough to strike fear into the heart of any mariner.

British charts were no better. The map publisher Thomas Jefferys was equally appalled by the general insensitivity British mariners showed towards precision. "The generality of mariners," he wrote in 1755, "seem to know of no quality in observing latitudes, farther than to find the place where they are bound to; and when they come in sight of land, lay the quadrant aside, as an instrument no longer of use, and sail by direction of the coast." While they might be useful for finding a general route from one coast to another, their mapping of inshore waters—where they were needed the most—was very unreliable. Either the chart's scale was too small

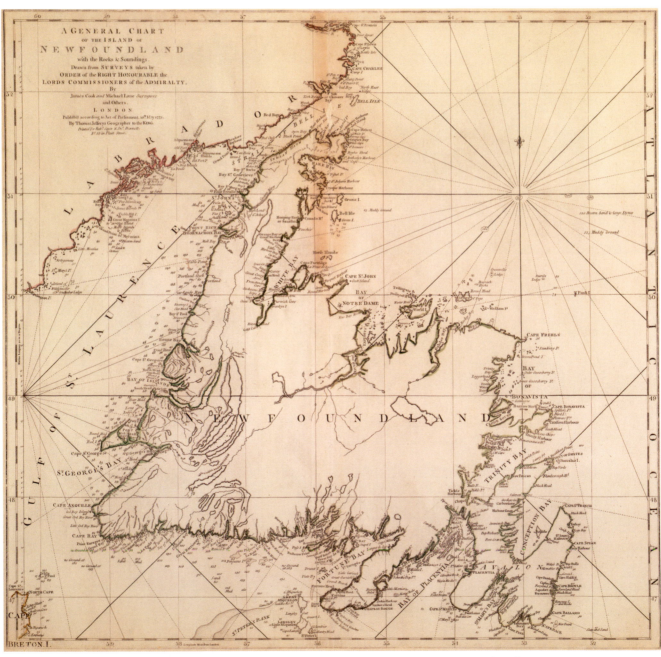

The survey of Newfoundland waters by James Cook and Michael Lane provided the British Admiralty with its first accurate charts of the island's coast. Their survey was published in this chart by Robert Sayer and John Bennett of London in 1775.

James Cook and Michael Lane, *A general chart of the island of Newfoundland with the rocks and soundings,* London, Thomas Jefferys, 1775. From *The North-American pilot for Newfoundland, Labradore, the Gulf and River St. Lawrence ...,* vol. 1, 1777, NMC 52302

or it lacked vital detail on such matters as variations in the lie of the land, the nature and size of shoals and rocks, and the best channels for entering a harbour. "Sailors dare not rely on them in times of danger and difficulty," admitted the British hydrographer Murdoch Mackenzie in 1755. Mariners of the day, he continued, considered "an error of 8 or 10 minutes but a trifle."

To move their fleet up the St. Lawrence, the British called on its younger officers to exercise their surveying skills, but it was James Cook (who would later distinguish himself in the South Pacific) and Joseph F. W. Des Barres who took on the bulk of the responsibility. Working under constant harassment from French guns, their surveys of the shoals and channels in the St. Lawrence gave the British an immediate advantage that was the envy of the French governor. "The enemy [the British] have passed 60 ships of war where we dare not risk a vessel of 100 tons by night or day," announced the nervous governor just before the British attack on Québec.

Once Britain gained a foothold in New France, making new maps and charts of the colony was an immediate priority. Since it was always possible that later treaty negotiations could see the region returned to France, this might be their only chance. In the end, the Treaty of Paris, which was signed in 1763, heralded the end of the Seven Years War and did indeed hand back a part of the former colony to France: the small islands of St. Pierre and Miquelon. For the French, the two rocky islands were a prize acquisition as they were situated in the heart of a lucrative gulf fishery. When the French governor-designate, M. d'Anjac, and a company of soldiers and settlers arrived to retake possession of the islands in the spring of 1764, the British were still busy surveying them. Such was the importance placed on accurate charts that Capt. Charles Douglas of the Royal Navy was willing to jeopardize Britain's shaky relations with France by continuing the survey. In a diplomatic coup that has never been completely explained, he proceeded to procure "all the time I could" for the surveyor, James Cook. After a series of contrived delays, the islands were finally handed over to a highly enraged d'Anjac on July 31 and James Cook presented the Admiralty with a set of incredibly detailed charts scaled at 1:221,760 (1 cm to approximately 2.2 km).

With its title to the former French colony now reasonably secure, the British turned their attention to

three projects that would supplement Murray's land survey of the St. Lawrence already underway (chapter 2). Two were initiated by the British Admiralty: a survey of the waters off the coasts of Newfoundland and Labrador by Michael Lane and James Cook (1763–75) and a survey of the Nova Scotia coast by Des Barres (1764–75). The other project was undertaken at the suggestion of the Board of Trade—an advisory agency to the British government that helped it to manage colonial affairs and to make the American colonies profitable to the Crown. Because of their importance to the fisheries, it included a survey of Prince Edward Island, Îles-de-la-Madeleine, and Cape Breton Island (in that order) by Samuel Holland (1764–76).

Luckily for the British, their need for up-to-date charts of Canada's Atlantic coast coincided with a

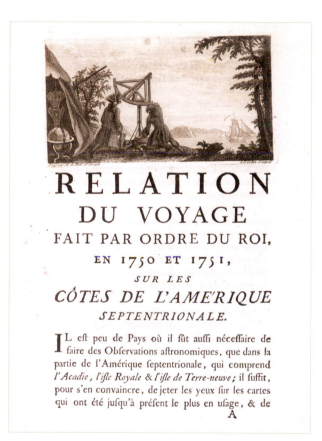

In this 18th-century engraving, the French hydrographer the Marquis de Chabert, a contemporary of Des Barres, uses a quadrant to calculate his longitude and latitude during a survey of the coast of New France. The instrument was named after its arc, which formed a quarter of a circle, and was used to measure the angle between the sun's noontime zenith and the horizon.

Joseph Bernard, Marquis de Chabert, *Voyage fait par ordre du roi en 1750 et 1751 dans l'Amérique septentrionale: pour rectifier les cartes des côtes de l'Acadie, de l'Isle Royale & de l'Isle de Terre-Neuve, et pour en fixer les principaux points par des observations astronomiques*, Paris, De l'Imprimerie royale, 1753, title page

James Peachey's portrait of Joseph F.W. Des Barres was probably completed shortly after the publication of *The Atlantic Neptune*. In 1784 Des Barres's accomplishments were rewarded with an appointment as the first lieutenant governor of Cape Breton, then a separate British colony. Two years later he surveyed the settlement that would become Sydney. In 1804 he was promoted to governor of Prince Edward Island. Des Barres always felt that his work had never been adequately recognized, however, and up until his death in 1824, he had unsettled grievances against the Admiralty.

James Peachey, *Portrait of Joseph Frederick Wallet DesBarres*, pastel on laid paper, ca. 1785, C 135130

period that saw some remarkable advances in hydrography. The theodolite had been recently refined to make it an even more accurate instrument for measuring horizontal and vertical angles. And in the 1730s, the quadrant and the chronometer were invented—two instruments that would help in measuring longitude and latitude.

Seafarers had known for several centuries how to calculate their distance (or latitude) from the equator by measuring the height of celestial bodies above the horizon. East-west distances (or longitude) proved to be more troublesome, however. As early as 1474, the German astronomer Johannes Müller had developed a method for calculating longitude from the angular displacement of the moon from other celestial objects, but the method remained impractical for nearly three centuries because of the inaccuracy of existing lunar

tables. Fortunately, the Royal Observatory in Greenwich began collecting the required astronomical data and by 1766 had a complete set of reliable lunar tables published in the *Nautical Almanac*. Using these lunar tables, hydrographers could reduce the error rate in their longitudes to one nautical mile. However, the lunar distant method was not for everyone: the trigonometry was complicated and could take as much as three hours to complete for a single set of observations.

Traditional marine surveyors ran a traverse of the coastline. This method called for rough sketches of a coast to be made as a ship sailed a direct line off shore. By carefully noting the ship's course and by taking a succession of compass readings to prominent features, an experienced draftsman could record an outline of the shore, noting its islands, points, inlets, and harbours. Since the method was imprecise, it was taken for granted that there would be errors. Sitting in a ship offshore, a surveyor often found it difficult to determine whether an inlet was a bay, a strait, or a river mouth—or whether a stretch of shore was part of the mainland or an island. As well, it was not easy to fix the position of coastal features using a compass on a rolling ship's deck. One inaccurate reading could throw off a whole succession of readings.

Des Barres (also spelled DesBarres) was among Britain's new breed of surveyors who understood that a truly accurate marine survey must first begin on shore and must use a system of triangulation. He started with an iron chain of at least 350 fathoms (about 645 m) and a theodolite to establish a baseline along the shore. The end points of this line would be fixed, employing a quadrant to calculate their latitude and longitude. From this baseline he took a series of readings on distant points along the shore, the idea being to take as many readings as possible so that each point was measured from at least two other points. This procedure allowed Des Barres to set up a series of intersecting triangles and confirm the position of each measured point through the laws of plane trigonometry. As long as the length of one side and two angles of the triangle were known, the position of distant points, which normally would have been unmeasurable, could be accurately determined. Once all the points were plotted, a draftsman could then begin to sketch the shoreline between them.

Des Barres's rule of thumb was accuracy, more accuracy, and still more accuracy.

Readings on the depth of the ocean floor were taken by a lead weight attached to the end of a hemp line. The weight was dropped into the water and when it hit bottom, the depth of the ocean floor could be estimated (in fathoms) by the length of rope that had been let out. Sometimes a core of tallow was set into the weight. The sticky tallow picked up debris from the ocean floor, giving the mariners some indication of its consistency. The readings were taken from shallops manned by local Acadians, considered best suited to the job by Des Barres because of their familiarity with the type of boat used and with the inshore waters. The Admiralty thought otherwise, however, and

Des Barres ended up paying the labourers out of his own pocket.

Other than the Acadians, Des Barres seems to have had no permanent staff. From time to time he was able to recruit some junior officers as assistants, but the assignments were always made reluctantly and were temporary. No doubt the assistants worried that removing themselves from the watchful eyes of their commanders might lessen their chances for promotion. These young officers were well aware that survey work, no matter how useful it might be to the Admiralty, simply did not measure up to other duties for merit promotions. Des Barres himself was a prime example of the military's indifference to its surveyors. After ten years of dedicated service charting the waters

A hydrographer, possibly Des Barres himself, takes readings near the entrance to present-day Canso harbour, Nova Scotia. The surveyor's shallop, manned by Acadians, lies offshore. To guarantee accurate and reliable charts, Des Barres used the latest survey procedures available and often took a double set of readings.

Joseph Frederick Wallet Des Barres, *West shore of Richmond Isle, near the entrance of the Gut of Canso*, London, J.F.W. Des Barres, 1776. From *The Atlantic Neptune*, NMC 28131

Des Barres's chart of Nova Scotia and Sable Island was renowned for its quality and beauty. His soundings off Sable Island were in use for 75 years before a second survey was attempted in 1851 by Capt. Henry Bayfield of the Royal Navy.

Joseph Frederick Wallet Des Barres, *A chart of Nova Scotia*, London, J.F.W. Des Barres, 1784, NMC 90077

Des Barres maintained his headquarters at Castle Frederick (near present-day Windsor, Nova Scotia), shown in this contemporary watercolour. He and his assistants spent the winter months here transferring their survey notes into charts.

Joseph Frederick Wallet Des Barres, *Castle Frederick, Falmouth (Nova Scotia)*, watercolour over pencil on wove paper, ca. 1790s, C 008468

of Nova Scotia, he still held the low rank of lieutenant when he returned to England. Perhaps as an incentive to keep younger officers in his employ, Des Barres had a reputation for keeping "a good table"—an expense that he, again, paid for out of his own pocket, and that he later claimed to have cost him the significant sum of half a guinea a day.

The survey work was not without its hazards. The larger survey ships had limited manoeuvrability close to shore and always ran the risk of hitting uncharted rocks and shoals, the very features they were there to investigate. The smaller shallops, as Des Barres discovered, could easily swamp in the unpredictable conditions of the North Atlantic. His survey of Sable Island was accomplished under extremely hazardous conditions. Situated about ninety nautical miles off the southeast coast of Nova Scotia, the fog-bound, low-lying dunes that make up Sable Island lie directly across the shipping route between Europe and New England. The one-and-half-kilometre-long island had already claimed more than two hundred ships by the time Des Barres arrived to chart its shoals. "This survey the

Joseph Frederick Wallet Des Barres, *A View from the camp at the east end of the Naked Sand Hills, on the south east shore of the Isle of Sable*, London, J.F.W. Des Barres, 1779. From *The Atlantic Neptune*, NMC 108111

Des Barres's survey camp on Sable Island (above) and the harbour at Annapolis Royal (below) are two of 145 views that were incorporated into later editions of *The Atlantic Neptune*. Some of the views served no navigational purpose and may have been added by Des Barres in an effort to make the atlas more attractive to general buyers.

Joseph Frederick Wallet Des Barres, *Annapolis Royal*, London, J.F.W. Des Barres, 1781. From *The Atlantic Neptune*, C 002705

COPPERPLATE ENGRAVING

The intricate detail for which *The Atlantic Neptune* is famous was partially achieved by Des Barres through the intaglio printing process. Unlike woodblock printing, the intaglio printer engraved the image into a printing plate rather than leaving it in relief on the surface (chapter 1). During printing, ink was rubbed into the incised lines, and the surface wiped clean. When dampened paper was pressed hard against the plate, it penetrated the engraving and absorbed the ink. The pressure also left a faint outline of the plate edge—a plate mark—around the edge of the paper.

Most mapmakers in the eighteenth century, including Des Barres, preferred to use heavy sheets of copper as printing plates because the metal was stable yet soft enough to permit the engraving of fine detail. Later in the nineteenth century, steel partially replaced copper because it could be put through a process that hardened the metal and helped it to withstand wear. The steel plates rusted easily, however, making their storage for later print runs a challenge.

Copperplate engraving allowed for greater artistic expression than woodblocks: the plates were significantly larger, and the soft copper was ideal for making fine, smooth curved lines. With copper a skilled engraver could

Imprimerie en Taille Douce.

This view of an early intaglio print shop is probably not very different from the one Des Barres set up to print his charts for *The Atlantic Neptune*. To make his work more attractive to general buyers, Des Barres coloured many of his maps and drawings through a technique developed in 1768 by the French printmaker Jean-Baptiste Le Prince. The process called for parts of the copperplate to be covered with granulated resin or sugar and the plate exposed to an acid bath. The acid pitted the surface only in the interstices between the granules. The uneven surface allowed the plate to hold just enough coloured ink to make the resulting print appear as if it had been painted with watercolour, and hence, the process was called aquatint. By varying the strength of the acid and its exposure time, a variety of different coloured tones and textures could be achieved.

Jean Le Rond d'Alembert and Denis Diderot, *Encyclopédie, ou, Dictionnaire des sciences, des arts et des métiers ...*, Paris, Chez Briasson ..., David l'aîné ..., Le Breton ..., Durand ..., 1751–65, 17 vols.

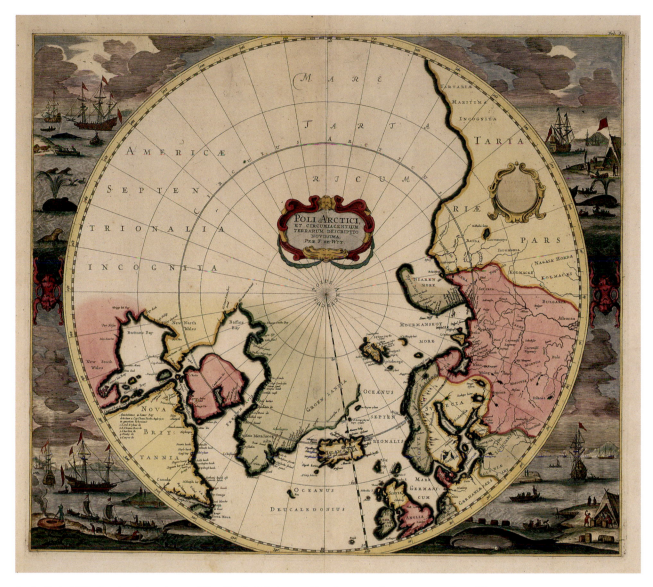

Frederik de Wit's hand-coloured map of the North Pole is considered among the most attractive of its time. The Arctic whaling scenes exhibit the high degree of artistic expression that cartographers had attained in the 18th century through copperplate engraving.

Frederik de Wit, *Poli arctici et circumiacentium terrarum descriptio novissima*, 1715, NMC 21059

easily vary the depth, width, and closeness of the lines, making it possible to add considerably more detail. Subtle shading and tonal effects were also achieved through crosshatching and stippling.

The greatest improvement copper offered, however, was its ability to allow changes. With care, parts of the map image and text could be hammered out and the plate re-engraved to correct mistakes or to add new information, making it unnecessary to start a whole new plate.

Many copper-plated maps were also hand-coloured using watercolours. Depending on the skill of the artist, the colouring could range from a crude wash to delicate shades. Because it was expensive—some printers employed women and children colourists to keep costs down—the process was undertaken only at the request of the customer. The premium charged for the service depended on the number of colours used and the extent to which they covered the map.

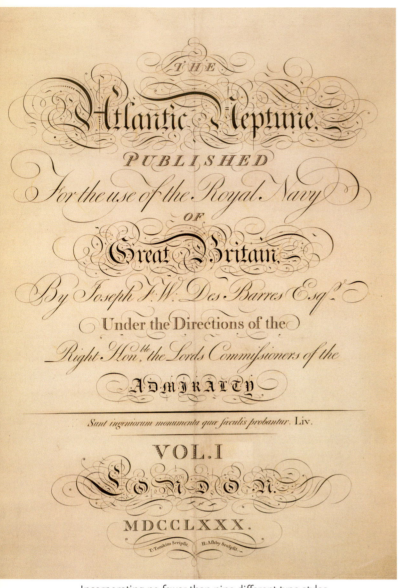

Incorporating no fewer than nine different type styles, Des Barres's *The Atlantic Neptune* had one of the most elaborate atlas title pages of the 18th century.

Joseph Frederick Wallet Des Barres, *The Atlantic Neptune*, London, J.F.W. Des Barres, 1780, NMC 27928

every intersected object, delineated the true shape of every head land, island, point, bay, rock above water, etc and every winding and irregularity of the coast; and, with boats sent around the shoals, rocks and breakers, determined from observations on shore, their position and extent, as perfectly as I could."

After ten years of surveying the waters of Nova Scotia, Des Barres returned to England in 1774 to prepare his charts for publication. Despite its heavy reliance on maritime trade, eighteenth-century Britain had no government agency responsible for chart production. The publication and distribution of hydrographic charts was completely in the hands of private industry.

Des Barres quickly assembled a staff of twenty to thirty-three compilers, engravers, and printers to begin the tedious task of transferring his drawings to copperplates. With a revolution looming in the American colonies, the Admiralty made it clear that it required charts of the entire eastern seaboard, not just Nova Scotia. Des Barres sought out the surveys of his colleagues Holland, Cook, and De Brahm and promptly began to incorporate them into a much larger marine atlas, which he titled *The Atlantic Neptune*.

The cost of engraving a copper plate in eighteenth-century Britain averaged from £2 to £30, depending on its size and complexity, but the Admiralty allowed Des Barres nearly £37. Although he left no indication about how much time was spent engraving each of the 247 plates used in his atlas, it was not unusual for some of his contemporaries to take as much as a year and a half to two years to complete a complicated plate. One historian has suggested that, with a complex map, an eighteenth-century engraver might be expected to produce only one square inch a day.

In just three years Des Barres was ready with the first edition of his atlas (1777), which was divided into five parts: Book I consisted of "impressions of all the charts [and] plates of my surveys of the coast and harbours of Nova Scotia, with a small book of tables of latitudes, longitudes, variations of the magnetic North, tides etc."; Book II contained "charts of the coast and harbours of New England composed, by command of Government, by various surveys, but principally from those taken by Major Holland under the directions of my Lords of Trade and Plantations"; Book III comprised charts of the Gulf of St. Lawrence, Cape Breton Island, and

Author has accomplished at the Hazard of his life," wrote Des Barres, "the surf upon the Shores admitting no boat to approach without the greatest risk."

Surveying the coastline was only the beginning: the data still had to be put on paper. During the winter months, Des Barres and his assistants drafted the season's work at his home, Castle Frederick, near present-day Windsor, Nova Scotia, then went back to the coast and compared the drawings to the actual landmarks, especially with regard to bearings. "I went along shore," wrote Des Barres in a letter to senior British command in 1765, "and re-examined the accuracy of

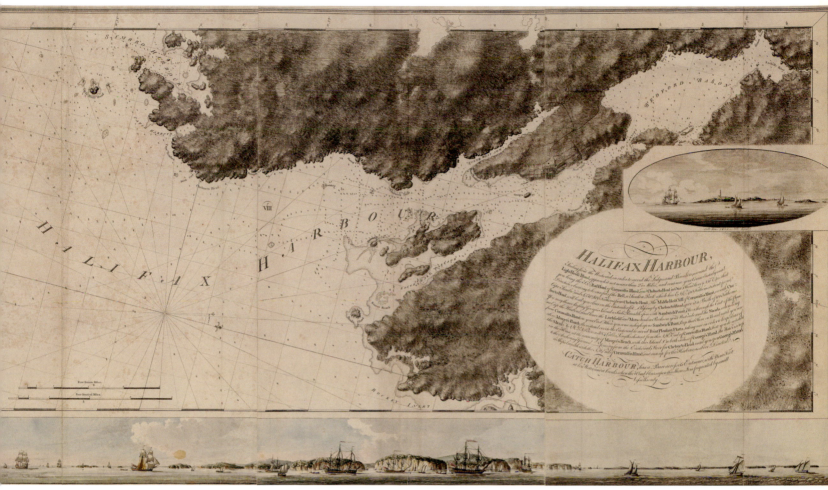

Modern-day estimates place the amount of copper used by Des Barres in *The Atlantic Neptune* at 2,100 kilograms. Each plate, including this one of Halifax harbour, was probably good for about a thousand prints before wear from the intense pressure of the press deteriorated the image to the point where it would have to be reworked, adding yet another production expense.

Joseph Frederick Wallet Des Barres, *Halifax Harbour,* London, J.F.W. Des Barres, 1776, NMC 28060

Prince Edward Island; Book IV, the coast of North America south of New York; and Book V, "various views of the North American coast."

In later editions of the atlas (published in 1780, 1781, and 1784), the views in Book V were incorporated into the appropriate first four volumes. Traditionally, mariners used views of the shoreline as an aid in helping them find their position along a coast. But Des Barres's colourful depictions were much more artistic, incorporating harbour scenes with ships at anchor and people involved in daily routines. Perhaps these images were included in the atlas with an eye to attracting sales from the general public, not only mariners.

Praise for *The Atlantic Neptune* came from all corners of Europe. A reviewer of the 1784 edition published in *L'Esprit des Journaux,* Paris, called it "a superb Atlas … indispensable for the navy … of the highest degree of beauty and superior to everything of the kind that has heretofore been published … of the highest possible utility for navigation and commerce."

The greatest tribute, however, was in knowing that his charts were helping to make eastern waters safe. Capt. Hyde Parker of the *Phoenix* was among those grateful for their existence. Having been tossed about in a storm for three weeks in 1778 and unable to make astronomical observations, the ship's officers believed they were somewhere off Cape Cod. After taking a few soundings and comparing their readings to Des Barres's charts, they realized the *Phoenix* was quickly approaching the dreaded Sable Island. A change of course saved the ship and crew from certain disaster. ❧

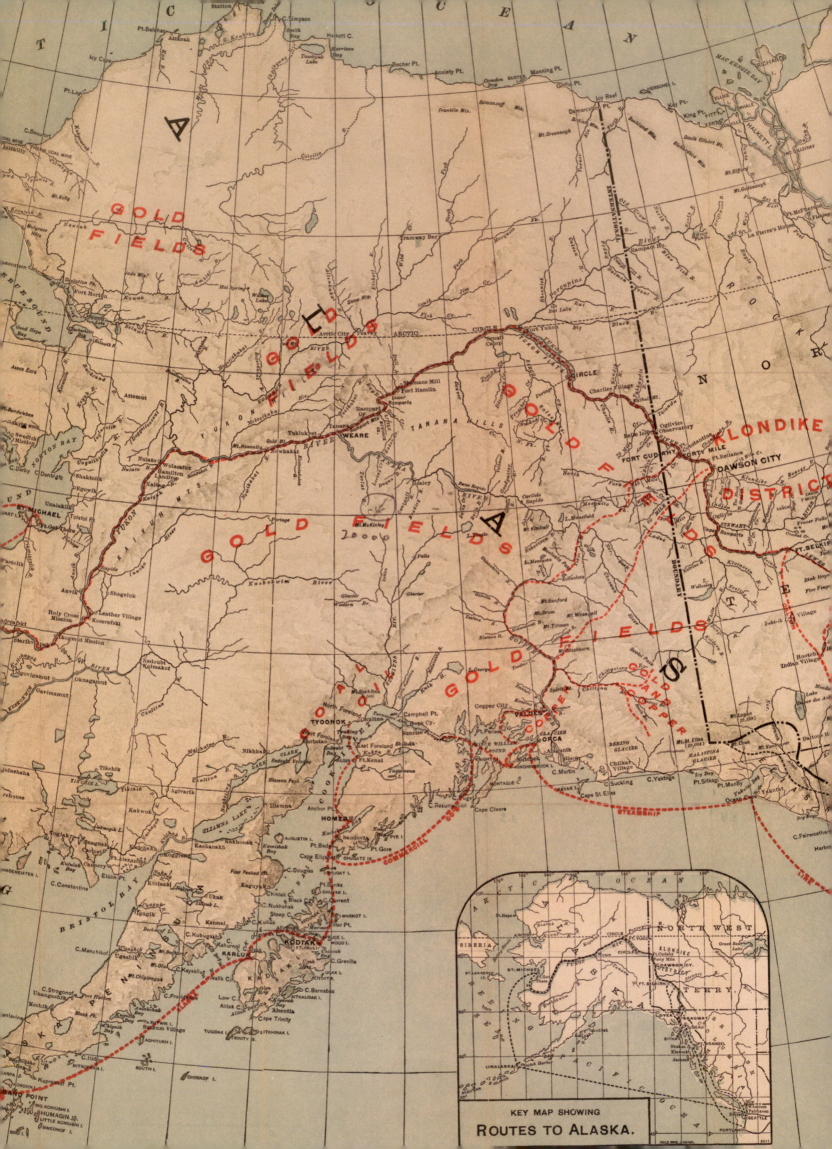

KEY MAP SHOWING
ROUTES TO ALASKA.

CHAPTER 8

Going for Klondike Gold

"THERE ARE MORE WAYS of making money than by going to the Klondike," declared Tappan Adney, the special correspondent for the *London Chronicle* and *Harper's Weekly*, who was anxiously hoping to find transport for his year's worth of provisions and mining equipment. Although the fees for packing goods across the Alaska Panhandle to the headwaters of the Yukon River at Lake Bennett were grossly inflated, Adney reluctantly handed over his money. If he could get his entire outfit to Lake Bennett in a few weeks with help from professional packers, there was a good chance he could make it all the way to the Dawson City goldfields before winter freeze-up. Although news of the gold strike on Bonanza Creek in the Klondike was barely a month old, there were already several thousand fortune seekers competing with Adney in a mad scramble to reach Lake Bennett. Everyone was heading for what the southern press had widely hailed as the largest gold find in history.

The Dawson City strike made headlines around the world when, in late July 1897, the steamer *Portland* docked in Seattle carrying a handful of weather-beaten miners and some two tons in Yukon gold. The arrival of a treasure ship from the north caught the North American continent still reeling from the effects of a decade-long depression. Some welcomed news of the gold strike as a final chance to bring financial dignity to their lives; others saw it as their last opportunity for a frontier adventure. Whatever their reasons, in less than twenty-four hours the Klondike gold rush was

The two packers in the foreground belonged to one of several Aboriginal groups that occupied the area travelled by the stampeders, but their communities were never shown on any of the maps produced for the Klondike trade. Indeed, Klondike maps would seem to suggest that the landscape was largely unoccupied, and hence, ready for exploiting.

La Roche, Seattle (firm), *Packers on Dyea Trail, near Stone House, Alaska,* ca. 1897, C 028645

In reference to its sudden rise to fame and seemingly unlimited potential, Dawson City was nicknamed "the Chicago of the north" at the peak of the gold rush in 1899.

Unknown photographer, *Aerial view of part of Dawson City*, ca. 1897, C 000675

on. Over the next eighteen months, an estimated 100,000 gold seekers set off for the Yukon in a stampede that would affect almost every community in western Canada and the United States.

Within a matter of days, thousands of stampeders began pouring into western "gateway" communities—San Francisco, Seattle, Portland, Vancouver, Victoria, Calgary, and Edmonton. Their demand for transportation to the north, and for supplies to carry them through a year of prospecting, created a marketing paradise for local merchants. In Seattle the adventurers brought in an extra $325,000 to the local economy in the first month alone. With fortune hunters from around the world arriving at their doorstep, it did not take long for western merchants to realize, just as Adney had, that the real money to be made in the Klondike gold rush was from the stampeders themselves, not from the goldfields.

Map publishers, in particular, found the influx of gold seekers to be good for business. Since so little was known in the south about this remote corner of the continent, the fortune hunters were naturally desperate

The stampeders' tent camp on Lake Bennett was the terminus of the extremely arduous Chilkoot and White Pass trails. Dawson City was about 925 kilometres downstream.

Eric A. Hegg, *Lake Bennett, B.C.*, ca. 1897, C 008258

for any information they could find, and maps were particularly high on their lists. The Province Publishing Company of Victoria and Vancouver, British Columbia, was among the first to hit the newsstands along Canada's west coast with a map of the Klondike. Their four-colour wall map of what is today Alaska, Yukon Territory, and northern British Columbia was produced in three formats: a paper edition that sold for 50 cents, a mounted edition that sold for 75 cents, and a water-proof edition that sold for a dollar. None of these prices was insignificant considering that a night in a Vancouver hotel cost about a dollar and a meal about 20 cents. Despite its high price, the map quickly went into a second, revised printing of 100,000 copies, and eventually a third edition.

Apparently, some stampeders were willing to pay almost anything for a map of the goldfields, particularly if they felt it gave them an advantage over their fellow treasure hunters. One group, met by Charles Camsell on the trail out of Edmonton, admitted to paying $300 for their sketch map of the gold-bearing creeks in the Yukon. The son of a northern Hudson's Bay Company factor, Camsell had spent all his life in the north and was initially shocked by the naiveté of these greenhorns. Before long he found several other prospectors with copies of the same map, which "the vendor must have sold to dozens of good seekers for the same amount." "In one case," continued Camsell, "the owner of the map told me he had got it from a friend while the two of them were engaged in shingling a barn in the Okanogan country in the State of Washington. The maps were no good as these fellows found out later, and the person who sold them was probably never at the locality in his life."

Not all Klondike maps were intended to be sold to unsuspecting would-be prospectors. Some were distributed by the major outfitters and transportation companies as a way of publicizing the various services they offered, and so were distributed free of charge. The Seattle firm of MacDougall and Southwick, for example, put together a map brochure in which they presented themselves as a reliable company with knowledge and experience on Yukon matters. The brochure featured a map of Alaska and the Yukon on one side; the reverse offered pertinent information on Canada's placer mining regulations and tariff laws.

As suggested by this advertisement, maps of the Yukon were popular money-making ventures in their own right throughout the Klondike stampede.
The Calgary Daily Herald, January 6, 1898, p. 2

Together, map and text supported the outfitter's claim as a company with concern for the stampeder's success in the goldfields.

Western merchants with a vested economic interest in the gold rush found maps to be particularly useful in maintaining the public's furore for the Klondike. On the pretext of providing geographical information, they often manipulated the map image so that it complemented the public's vision of the region's seemingly unlimited possibilities of wealth. For example, publishers of Klondike maps almost inevitably shaded the Alaska and Yukon land masses in yellow, no doubt a crude attempt at a subliminal association with the goldfields they were promoting. To make the association even more apparent, the publishers frequently scattered labels reading "gold fields" across the vast unmapped regions of the northwest. Usually these labels were in a contrasting colour to make them stand out. As well, the "gold fields" labels were often spaced

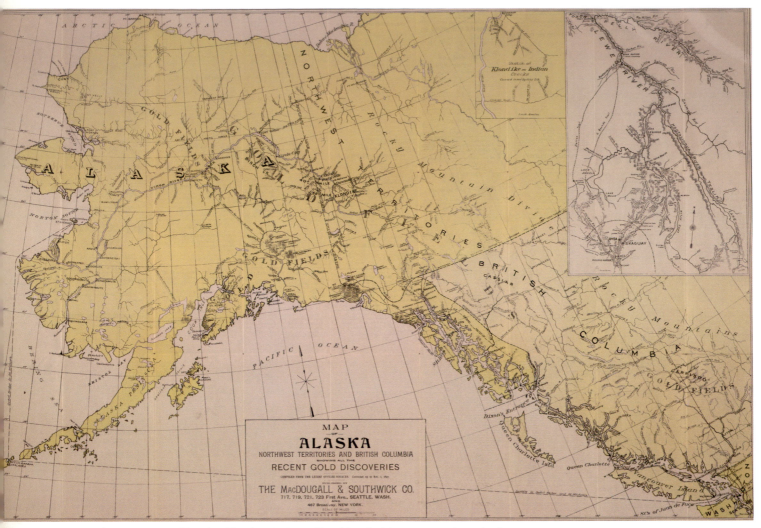

Like most maps of the Klondike, the MacDougall and Southwick map was coloured yellow, no doubt an attempt at a subliminal association with the goldfields it was promoting.

The MacDougall and Southwick Co., *Map of Alaska, Northwest Territories and British Columbia showing all the recent gold discoveries: Compiled from the latest official sources*, Seattle, MacDougall and Southwick Co., 1897, NMC 98202

so that they covered most of the Alaska and Yukon interior, even the areas that lay well beyond Dawson City. The combined effect of colour and labels provided the mapmakers with a clever visual metaphor for the widely held notion propagated by the southern promoters that the Klondike was just one river in a vast territory containing hundreds and thousands of similar waterways. If the Dawson City region could provide such a rich treasure, so the reasoning went, then the possibilities of finding gold in the northwest and all its waterways were almost beyond comprehension.

Some maps produced by private firms also labelled earlier and much smaller gold strikes, such as those that took place in the Omineca, Cassiar, and Cariboo regions of British Columbia's interior. Interestingly, all

these labels were usually in the same typeface and colour as the one used for the Dawson City discovery. This technique forced the reader to ignore each region's uniqueness and to assume incorrectly that all the finds were as important as the Klondike strike. Once again, the images served the promoters of the Klondike stampede by providing them with further visual evidence to support their sensationalized reports on the vast mineral potential of the northern frontier.

Perhaps the most blatant attempt to manipulate the cartographic image was the extent to which mapmakers exaggerated the accessibility of the Yukon from the south. Whether promoters wanted to champion the interior "all-Canadian" routes from Edmonton and Calgary, or the Pacific coastal routes from Vancouver,

Victoria, Seattle, and San Francisco, more often than not they were all represented on the map in the same way as the major transportation corridors in the south. In fact, it was not unusual for promoters of a particular route to mark the trail even if it existed only on paper and could prove life-threatening to novices of wilderness travel. Without exception, the maps offered no warning to the Klondike-bound about the hardships they could expect to encounter.

To many fortune seekers, the maps of the day seemed to suggest that the Lake Bennett route—the route taken by Tappan Adney—was the most direct and practical way to the Yukon goldfields, offering a combined overland and water passage from the Pacific coast, across the Chilkoot Mountains, to Lake Bennett, and the beginning of the Yukon River.

Maps supporting the route showed the overland portion to measure about 42 kilometres. What the maps conveniently neglected to mention was that each stampeder would be required to make about twenty trips (each way) across this narrow land strip, either through the 1,097-metre-high Chilkoot Pass or the 793-metre-high White Pass. With each trip, the stampeder would be required to carry (or pay someone to carry, as Adney did) a pack weighing 45 kilograms if he wanted to relay his entire prospector's outfit to Lake Bennett. This one oversight added, on average, almost a thousand kilometres to the trip and, depending on the weather, would require an extra five to eight weeks of travel. Even with all the help Tappan Adney received, it took him a total of ninety-two days to reach Dawson City, nearly two and a half months longer than the

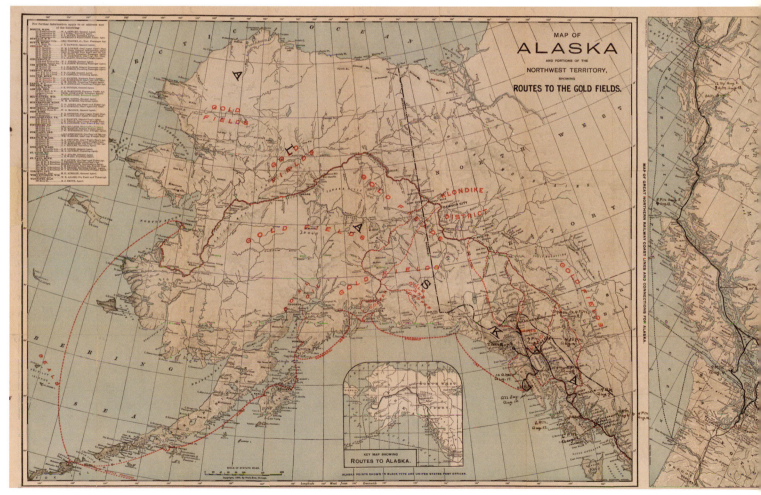

The optimism southern promoters of the Klondike stampede felt about the Yukon is clearly expressed in this map where most of the interior has been divided into goldfields, even those areas that were unexplored and unmapped. The word "gold" appears eight times in bold red letters.

Poole Bros., *Map of Alaska and portions of the Northwest Territories showing routes to the gold fields,* Chicago, Poole Bros., 1898, NMC 78696

Rustic conditions on the Klondike trail caused many hopefuls to return home without ever having seen either Dawson City or the goldfields.

La Roche, Seattle (firm), *A packer's home foot of canyon, Dyea Trail*, 1897, C 028647

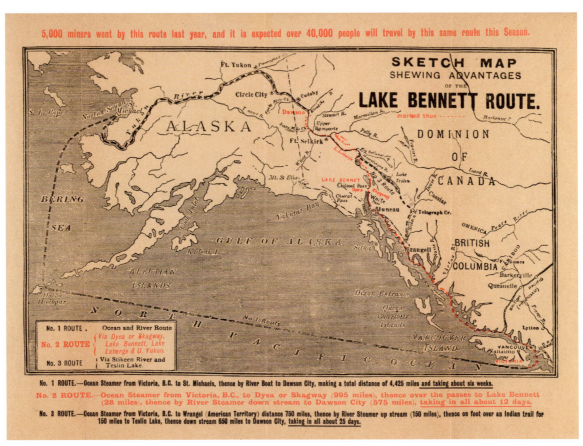

"Hell could not be worse than this," wrote E.H. Wells as he tackled the White Pass trail to Lake Bennett. "The opening of the White Pass ... was not a blunder—it was a crime," echoed Tappan Adney over this widely touted route to the Klondike. Yet despite such unequivocal denouncements, promoters of the Lake Bennett route still circulated map advertisements declaring the route the quickest to the Klondike goldfields.

Unknown cartographer, *Sketch map shewing advantages of the Lake Bennett Route*, 1897, NMC 13240

Stampeders relay their "ton" of food and equipment over the summit of the Chilkoot Pass. They were required to make at least 20 such trips across the 1,097-metre-high pass carrying packs weighing 45 kilograms. Although this back-breaking task added more than 1,000 kilometres to their journey, it was never mentioned in any of the maps promoting the route.

J.G. McJury, *Digging out after snowslide (at Chilkoot Pass, B.C.)*, 1898, C 026413

twelve days optimistically claimed by at least one map promoting the route. Not surprisingly, many who tried this route to the goldfields turned back disillusioned and broken. As Adney himself pointed out, of those who continued, many "regret having started" and "all of them blame the misrepresentation about the trail."

Given the types of maps circulated, it is perhaps to be expected that many fortune seekers were terribly optimistic about how easy it would be to get to this fringe of the frontier. Alphonso Waterer, a stampeder from southern Alberta, came across just such a group in front of an Edmonton hotel. "A bunch of men were discussing the Yukon, and were most anxious to learn where we came from and how we proposed reaching Alaska," he recalls in his unpublished memoirs. "They were a motley crowd and most of them rather too inquisitive from our point of view. One party of five were looking over a map and as they thought that Dawson appeared to be only a

THE FEDERAL GOVERNMENT STEPS IN

The Klondike gold rush caught the federal government completely unprepared. News of the gold strike brought thousands of inquiries from around the world for up-to-date information on the region. Inundated with requests for maps, the major federal mapping agencies—the Geological Survey of Canada and the Department of the Interior's offices of the Surveyor General and Chief Geographer—soon found their decade-old map supply exhausted.

The Department of the Interior quickly set out to make up the deficiency through the publication of two new major maps, of which the most ambitious was the *Yukon Map*. Consisting of ten sheets at a scale of 1:380,160 (1 cm to approximately 3.8 km), it covered much of the highly contentious Yukon-Alaska border from the mouth of the Stikine River in the south to Dawson City in the north. The map was compiled from the photo-topographic surveys of the International Boundary Commission (chapter 11), and "any other surveys and explorations that were available and authentic." Wherever possible, the three-coloured map was contoured and the hills shaded to provide a realistic impression of the mountains and passes. In order to expedite production, the government made use of all the lithographic print shops in the Ottawa-Montréal region. They produced 5,200 copies of each sheet at an average cost of $150 per sheet.

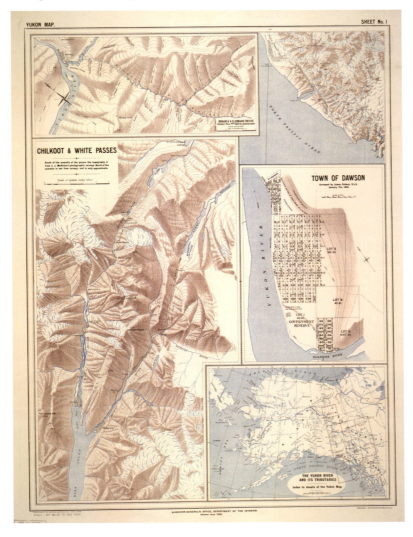

Chief Geographer J. Johnston began a second map that showed, at a reduced scale, the entire northwestern portion of the continent, from Edmonton to the Arctic coast and west to central Alaska. Working night and day, Johnston drove himself to the point of exhaustion and died before the map was finished, leaving it to be completed by his successor, Jacob Smith. A preliminary edition of the map was published in January 1898; a second edition, in a larger format and with more detail, followed the next year. "It is a very valuable map and includes the results of all the established surveys up to '98," announced the *Edmonton Bulletin* in

Published in June 1898 by the Surveyor General, the *Yukon Map* covered a 518,000-square-kilometre section of southwestern Yukon and adjacent British Columbia, and for the first time detailed the Chilkoot and White passes, Dawson City, and the claims staked along Eldorado and Bonanza creeks.

Canada, Office of the Surveyor General, *Yukon Map*, Ottawa, Surveyor-General's Office, 1898, sheet no. 1, NMC 27874

During the Klondike gold rush, draftsmen in the federal government's mapping unit of the Department of the Interior worked round the clock to bring reliable, up-to-date maps of the Yukon to the public.

Canada, Dept. of the Interior, *The Drafting Division, Natural Resources Intelligence Branch, Ottawa*, ca. 1920, PA 043739

October 1899. The map was based on surveys by the Department's Topographical Branch, the International Boundary Commission, the Geological Survey of Canada, and the British Admiralty. Despite this impressive list of credits, there were still large areas of the northwestern interior left blank and designated as "unexplored." The map was copied extensively by private publishers, but in many instances the "unexplored" label was changed to read "gold fields."

The preliminary edition of the federal government's new map of northwestern Canada cost Chief Geographer Johnston his life. Unfortunately, by the time the map was available, the great rush of would-be prospectors to the Yukon was largely finished.

Canada, Office of the Surveyor General, *Map of the north-western part of the Dominion of Canada*, preliminary ed., Ottawa, Mortimer Co. Litho., 1898, NMC 11623

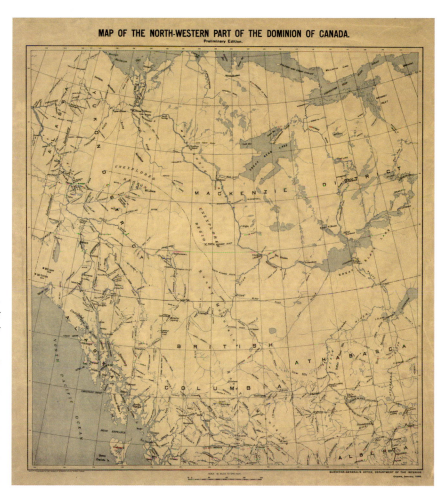

MAP OF THE NORTH-WESTERN PART OF THE DOMINION OF CANADA.
Preliminary Edition.

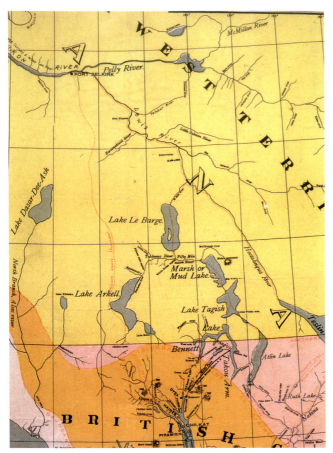

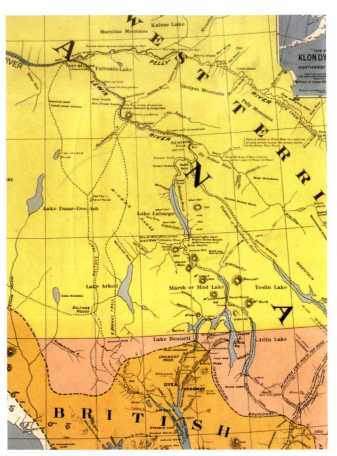

The Province Publishing Company, *"The Province" map of the Klondyke and the Canadian Yukon and routes thereto* (detail), Victoria, Province Publishing Company, 1897, NMC 44324

The Province Publishing Company, *"The Province" map of the Klondyke and the gold fields of the North-West Territory* (detail), Victoria, Province Publishing Company, 1897, NMC 17626

Although Klondike maps could be deceptive, mapmakers at least attempted to include the latest geographical information where it suited them. These details from the first (left) and second (right) editions of the Province Publishing Company's map of the Klondike show the extent to which the headwaters of the Yukon River were revised in the second printing.

short distance away they made up their minds to reach it in 2 or 3 days." In reality many of those who followed the Edmonton route to Dawson City spent up to three years on the trail. Even the best equipped and most experienced adventurers took a minimum of fourteen months.

With so many southern promoters shamelessly plugging the unlimited potential of the Yukon and its accessibility, it is no wonder Tappan Adney concluded that "the most important item on the list [of supplies] is good advice—plenty of it." A few days later, when he saw the situation on the Lake Bennett trail for himself, he came to the sickening realization that finding good advice was easier said than done. "We have learned already to place no reliance upon any person's word," confessed a more experienced and cynical Adney.

Fortunately for Adney, his efforts on the Chilkoot Trail paid off in one respect: at least he managed to reach Dawson City just ahead of winter freeze-up. A few days after his arrival, sub-zero temperatures took hold of the land, effectively shutting it down until the following spring. But in the end, his efforts turned out to be completely in vain. All mining claims on all the gold-bearing creeks were already staked and had been for about a year.

The Klondike gold rush proved to be a bust for most stampeders, including Adney. Southern outfitters and transportation companies were the only ones who really profited. Although exact figures are difficult to come by, it is estimated that the stampeders brought some 14 million kilograms of solid food into Dawson City over a two-year period, and probably an

A map published by the Canadian Pacific Railway in 1898 shows the railway connecting, either directly or indirectly, with all known routes to the goldfields. It implicitly suggests that stampeders could travel on CPR trains no matter which route they took to the Yukon. The map even includes routes through the interior of Alberta and British Columbia, which would have been madness for anyone to attempt.

Canadian Pacific Railway, *Map of Canadian Pacific Railway and connections showing routes to the Yukon gold fields Alaska, Klondike and the northwestern mining territories of Canada*, Chicago, Poole Bros., 1898, NMC 8447NMC 008447

equivalent amount in camping gear and mining equipment. All of these goods were purchased in the south and were shipped to northern destinations by southern transportation companies.

Maps were particularly useful to those southern merchants looking to cash in on this marketing bonanza as they helped the promoters inspire visions of wealth among a public facing uncertain and difficult economic times and, by doing so, conveniently added to the magical mania of gold rush rhetoric. Brilliantly deceptive and misleading, Klondike maps were powerful tools of persuasion. ❧

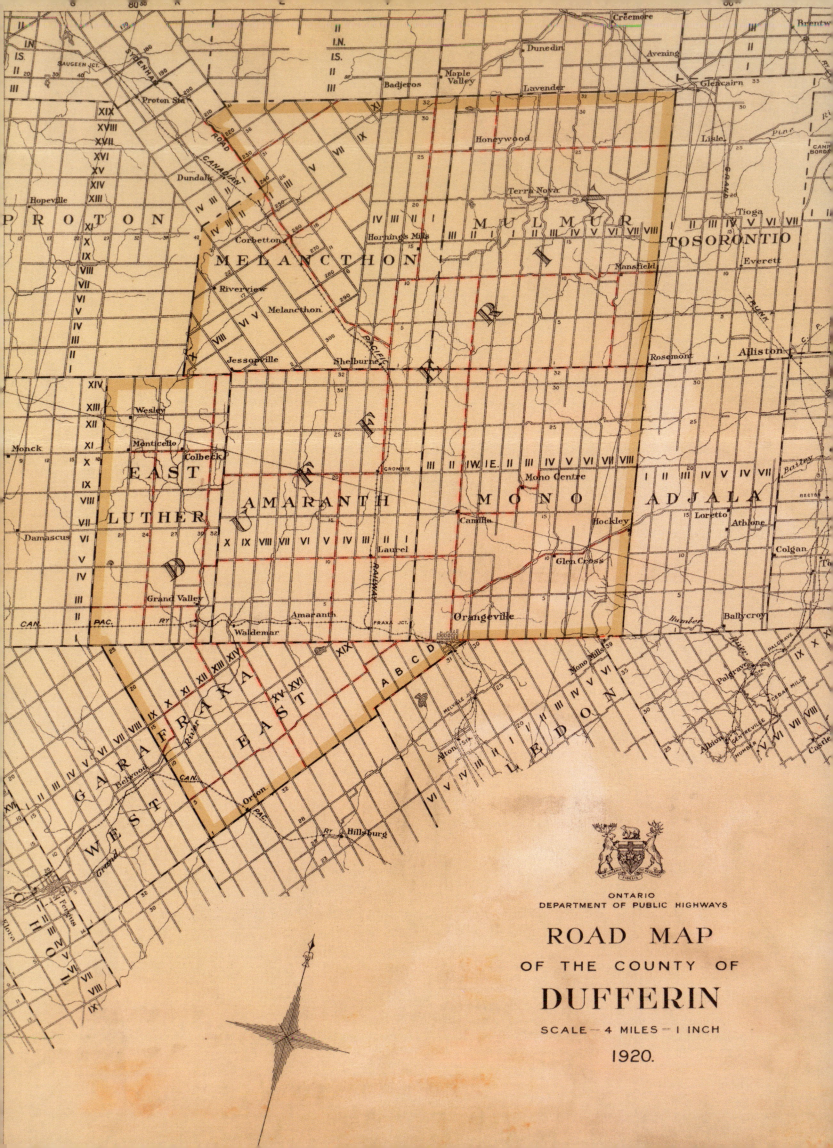

ONTARIO
DEPARTMENT OF PUBLIC HIGHWAYS

ROAD MAP
OF THE COUNTY OF
DUFFERIN

SCALE — 4 MILES — 1 INCH

1920.

CHAPTER 9

Maps for Your Motoring Pleasure

"WITH THE EXCEPTION OF A FEW DAYS, the months of October and November are our best motoring months and Western Ontario is fortunate in having fairly good roads," announced the *Toronto Daily Star's* automobile editor in the fall of 1916. While Canada's youth faced the muddy trenches of the Somme, the motoring clubs back home debated the best weekend tours for their shiny new McLaughlin Buicks, Stanley Steamers, and Model T Fords. By most accounts, the recently completed concrete highway between Toronto and Hamilton—Ontario's first—meant that the best roads in the province were centred in the Hamilton area and the city had become a favourite destination for many weekend Toronto motorists.

What made the Hamilton tour particularly pleasing was the excellent accommodation afforded by the city's new Royal Connaught Hotel. The hotel's club breakfast, luncheon, and table d'hôte dinner were widely reputed to be the finest in the country, and "make Hamilton for dinner" had become a rallying cry of

The thrill of the open road brought these tourists to Banff National Park in the mid 1920s. The free road maps distributed by oil companies and all levels of government helped to promote Canada's budding touring industry to a growing number of automobile owners.

William J. Oliver, *View of Banff from Tunnel Mountain*, 1929, PA 057241

Toronto drivers. Even better, the Royal Connaught had issued a booklet with a road map giving easy directions to the hotel, and it was willing to send the map, free of charge, to anyone on written application. "There is probably no other place in Ontario where motorists are so well served," concluded the *Star*'s editor.

By the time the Royal Connaught issued its road map in 1916, auto touring was already one of the fastest growing industries in the country. Although it had been only eighteen years since the first gasoline-powered car entered Canada—an American-built Winton purchased by John Moodie of Hamilton—there were already sixteen motor clubs in the country and some 80,000 registered automobiles. At the end of the Great War two years later, this number would jump to 277,000 vehicles, giving Canada the distinction of being second only to the United States in the num-

ber of cars per capita and third, after Great Britain, in total ownership.

It made good economic sense for the Royal Connaught Hotel to cater to the growing body of Ontario motorists. The increased number of affluent customers that the hotel could expect easily justified the cost of distributing a free map. Since most Ontario roads at the time were unmarked, the map not only encouraged motorists to make the trip, but once they were underway, it also helped ensure they would reach their destination.

The Royal Connaught's road map was not a completely new idea. Enthusiastic and determined cyclists had been touring the Ontario countryside since the mid 1890s and publishers had been selling them maps to navigate the maze of unmarked roads. As well, the Ontario Motor League had been issuing its annual

This 1895 map of the Hamilton and Niagara region was designed especially for cyclists planning a weekend jaunt in the country. It showed only those roads that were considered passable. The railway lines were important for cyclists wanting to ride the rails in one direction.

D.B. Street, *Cyclists Road Map, showing all the main travelled roads, towns, villages etc. between Toronto and London including the Niagara District*, 1895, NMC 43015

Improved printing techniques have enhanced the overall communication power of road maps. Today's mapmakers have a wide range of lines, typefaces, colouring, shading, and symbols at their disposal for conveying information in greater detail.

H.K. Carruthers, *Official road map of eastern Ontario and part of Quebec, New York and Vermont, showing provincial, state, county and township highways*, 1923, NMC 21548

Official Automobile Road Guide of Canada since 1908. The *Guide*, priced at $4, included maps and detailed written descriptions of dozens of motor tours (complete with odometer readings), all of which had been tested by League members. Some of the tours even included photographs of unmarked intersections indicating required turns.

Once the passenger compartment of automobiles was enclosed, the impracticality of guidebooks started to become obvious. By the late 1920s they were becoming increasingly cumbersome and expensive to produce: the Ontario Motor League's guidebook had grown to 400 pages and Scarborough's Automobile Green Book reached more than 900 pages. Despite their bulk, with the number of new roads being developed each year, the guidebooks simply could not provide detailed written descriptions for all possible routes available to motorists. In addition, the landmarks in the guides were not always current. "Turn at the white house" might be accurate one year but, if the

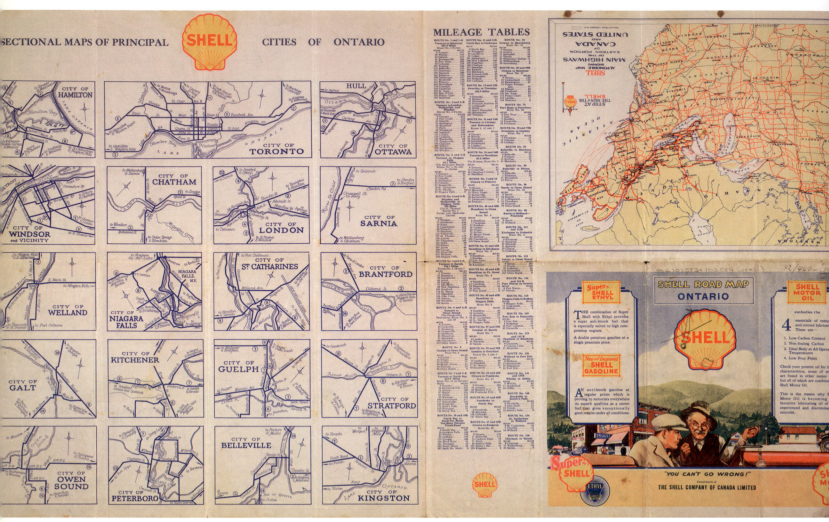

Along with free road maps came covers featuring friendly and attentive service attendants, willing allies of the confused motorists, as in this Ontario road map from the 1930s published by the Shell Oil Company of Canada.
Shell Oil Company, *Shell official road map of the province of Ontario*, 1930, NMC 103584

owners painted the house another colour, the direction could be misleading the next year.

What motorists needed was some other way to navigate country roads. The Ontario Motor League came up with a novel idea. Using a portion of its membership fees, it created its own yellow-and-black directional signs, which it tacked to telephone poles and trees. By 1915 the League had well over 13,000 signs in place, prompting the editor of *Canadian Motorists* to suggest that "there are few motorists who do any touring at all who are not familiar with the black and yellow signs of the League and who do not feel indebted to the OML for its work in this direction." Five years later, the League claimed that with a "road-sign car" dedicated to the task of "blazing trails," their familiar signs had grown to an estimated 40,000.

The League's idea caught on with some counties in the southern parts of the province. The Toronto and York Roads Commission, for example, placed its own signs at all intersections and jogs and assigned every road its own colour. These colours were painted on poles and trees and were edged with a white band to make them more easily visible at night. The Ontario Motor League used this system on the maps they printed so that motorists would correctly associate colours with certain roads. As well, the county had put in place a system of signs showing the mileage to the nearest towns. According to the *Globe*, the originator of the scheme, Mr. E.A. James, was working with other counties in the province to have the signs extended beyond York's borders in order to make the system more universal and useful to motorists.

By the second decade of the twentieth century, the automobile was firmly established as a permanent feature in the lifestyle of Canadians. Less and less an indulgence for the wealthy, it was becoming more and more a practical mode of family transportation. As the number of cars multiplied, so did the pressure on Ontario's lawmakers to build better roads. Although the province's Public Roads and Highways Commission dated back to the days before Confederation, serious highway construction did not get underway until 1915, when the Ontario legislature passed a Highways Act and created a new Department of Public Highways to administer it. Before 1915, road construction and maintenance was mainly a county responsibility. The result was an array of road conditions, most of them suitable for horse and wagon traffic but unacceptable for automobiles.

The new Act allowed the province to identify roads that would become part of a provincial system, made new money available for road upgrades, and set standards for road construction. For example, the Act required every main road to "have an allowance at least 66 feet [20 metres]" wide and made provisions for a Board of Road Trustees who had the authority to widen roads an additional 6 metres when it was considered desirable and to adopt more favourable routes if they felt it necessary.

Beginning in 1915 and continuing into the early 1920s, the Department of Public Highways compiled an inventory of the province's roads—a first for the province—and mapped this network on a county-by-county basis. The information was used to identify a highway system that linked most of the major centres in the southern part of the province. By 1923 the Department had widened several thousand miles of highway roads to 26 metres and had issued its first "official" road map—likely the first road map published by a provincial government in Canada. The map was printed in two colours (blue for lakes and rivers, black for everything else) and used three different line thicknesses to separate the provincial highways from county and township roads. Other than cities, towns, and villages—all of which were given equal consideration no matter how large or small—the map showed little else that might be of interest to motorists.

In an attempt to provide motorists with a more universal system of highway identification, the provincial government came up with the idea that, rather than colour its highways, it would number them. The simple logic of this scheme was immediately apparent, and the numbers were put in place in time for the opening of the 1925 tourist season. "There will [now] be no difficulty," declared the editor of the *Perth Courier*, "for the veriest amateur tourist finding his way around our Province."

Later editions of the province's official road map soon made up for the shortfalls of the first one. The 1926 edition, for example, showed not only highway numbers, but also identified fish and game areas. The province also made the map a little easier to read by using different colours for the various types of roads.

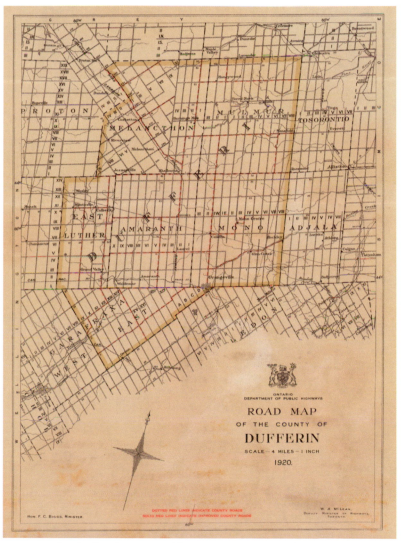

Beginning shortly after the First World War, Ontario's newly created Department of Highways surveyed and published road maps at the county level before issuing its first province-wide road map in 1922.

Ontario, Dept. of Highways, *Road map of the county of Dufferin*, Toronto, Dept. of Highways, 1920, NMC 3279

OFFSET PRINTING

The majority of road maps prepared for the Canadian market after the First World War were produced by offset lithography. In this technique, a flexible zinc plate (later aluminium) coated with a photosensitive material, with the surface treated to make it more porous, was mounted on a rotating cylinder. Exposure of the plate to an image hardened the coating on areas to be printed and washed away non-printing areas, which also rejected ink. The ink transferred from the plate to a rubber-coated blanket cylinder that imprinted or "offset" the image onto a sheet of paper. The rubber cylinder generally produced a better impression than printing directly from the metal plate, and offered greater flexibility by allowing maps to be printed on a wider range of media: wood, cloth, metal, leather, and rough paper. In offset printing, the printed image is neither raised above the surface of the printing plate (as in woodblock printing) nor sunk below it (as in copperplate printing).

The offset rotary press did not completely replace the flat-bed stone press for a number of years (chapter 5), but by the late 1930s, the versatility of the offset process led many printers and mapmakers to make the switch. Drawings could be produced and printed at almost any size and most line work and halftones could be accommodated. By 1926 a rotary press was in operation in the federal government's Surveyor General's office and by 1932 at least three such presses were in use.

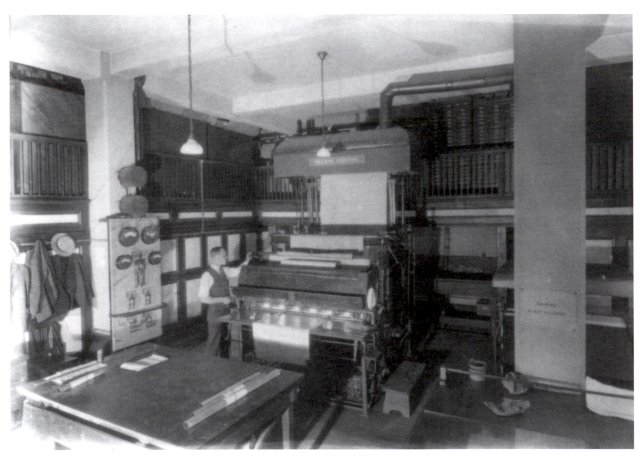

An offset printing press in the federal government's Natural Resources Intelligence Service around 1920, where many specialized maps for auto tourism were printed.

Canada, Dept. of the Interior, *Process printing, Natural Resources Intelligence Branch, Ottawa*, ca. 1920, PA 043743

A copperplate press typically printed about six to eight uncoloured copies an hour (chapter 7). A lithographic hand press was considerably faster at twenty to fifty copies an hour, and a mechanically driven high-speed offset press turned out several thousand coloured copies an hour. Offset printing also permitted the use of a greater variety of colours in fewer press runs and in exact register. The improved printing technique reduced map publishing costs to pennies a sheet and, for the first time, allowed their unlimited free distribution through neighbourhood stores.

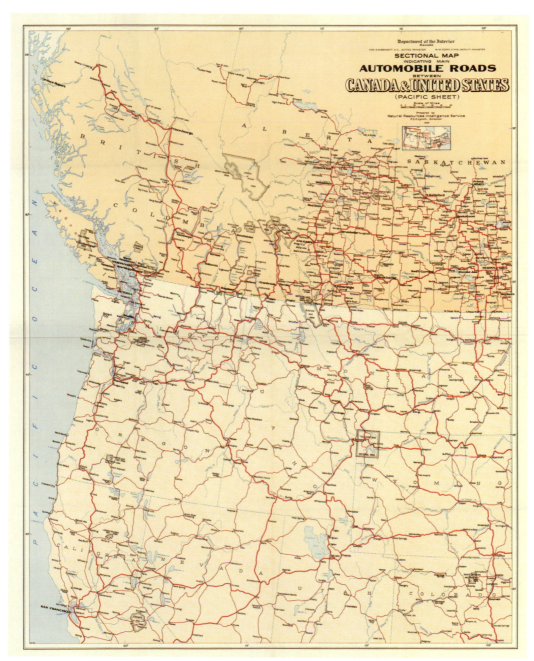

A series of four road maps published by the federal government's Natural Resources Intelligence Service in the mid to late 1920s focused on Canada's border areas with the United States. The series tried to demonstrate how easy it was for American motorists to cross into Canada.

Canada, Natural Resources Intelligence Service, *Map indicating main automobile roads between Canada and the United States*, Ottawa, Natural Resources Intelligence Service, 1926, NMC 25231

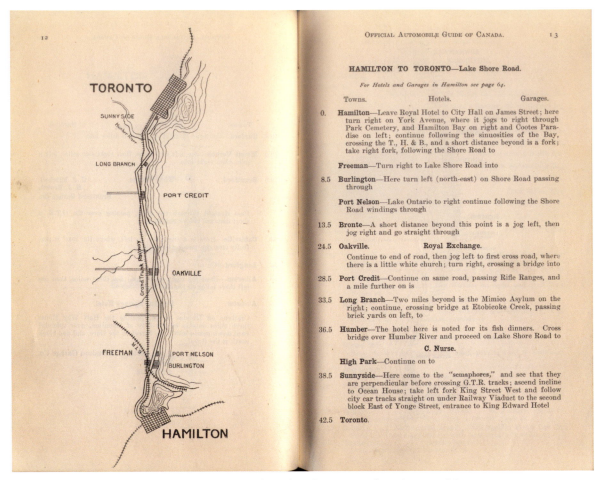

HAMILTON TO TORONTO—Lake Shore Road.

For Hotels and Garages in Hamilton see page 64.

Towns.	Hotels.	Garages.

0. **Hamilton**—Leave Royal Hotel to City Hall on James Street; here turn right on York Avenue, where it jogs to right through Park Cemetery, and Hamilton Bay on right and Cootes Paradise on left; continue following the sinuosities of the Bay, crossing the T., H. & B., and a short distance beyond is a fork; take right fork, following the Shore Road to

    Freeman—Turn right to Lake Shore Road into

8.5 **Burlington**—Here turn left (north-east) on Shore Road passing through

    Port Nelson—Lake Ontario to right continue following the Shore Road windings through

13.5 **Bronte**—A short distance beyond this point is a jog left, then jog right and go straight through

24.5 **Oakville.**   **Royal Exchange.**

    Continue to end of road, then jog left to first cross road, where there is a little white church; turn right, crossing a bridge into

28.5 **Port Credit**—Continue on same road, passing Rifle Ranges, and a mile further on is

33.5 **Long Branch**—Two miles beyond is the Mimico Asylum on the right; continue, crossing bridge at Etobicoke Creek, passing brick yards on left, to

36.5 **Humber**—The hotel here is noted for its fish dinners. Cross bridge over Humber River and proceed on Lake Shore Road to

    C. Nurse.

    High Park—Continue on to

38.5 **Sunnyside**—Here come to the "semaphores," and see that they are perpendicular before crossing G.T.R. tracks; ascend incline to Ocean House; take left fork King Street West and follow city car tracks straight on under Railway Viaduct to the second block East of Yonge Street, entrance to King Edward Hotel

42.5 **Toronto.**

Road guides featuring well-designed tours were popular with early motorists from the turn of the century to as late as the 1930s. As shown on this 1910 map of an auto tour from Hamilton to Toronto, the maps described where to turn ("where there is a little white church, turn left").

James Miln, *Official automobile road guide of Canada with map routes: authorized by the Ontario Motor League*, 4th ed., Toronto, Miln-Bingham, 1910, pp. 12–13

For the first time, it included an index to Ontario's cities, towns, villages, and lakes. Five years later, the map also sported references to the fifty-four historic sites designated by the Historical Sites and Monuments Board. The *Globe* greeted this last change with some enthusiasm; they thought that it would help visitors "who wish to learn something of the background of the country."

Not everyone was excited by the way road maps were helping motorists to open up the country to tourism, however. One visitor to the *Toronto Daily Star* offices in 1927 admitted that "in the old days" when auto touring was just beginning, automobiles were a good choice over the railway in finding the best fishing. But with everyone now using their cars to access these old fishing holes, the railway had once again become his best bet. "My rule for going fishing is to get a road map," announced the angler, "look for the places where there is most white space on it and then take the railroad into those regions."

The advent of balloon tires, safety glass, tail lamps, power windshield wipers, gas gauges, large luggage trunks, and road maps slowly turned Canada into a nation at ease behind the wheel of a car. By 1930 there were more than a million vehicles in the country. The automobile, now as much a part of household activities as the family pet, had transformed family vacations and Sunday afternoon drives into cross-country excursions.

Once small-town Canada made the connection between road maps, auto touring, and economic growth, local officials took it upon themselves to closely monitor road maps of all shapes and sizes to ensure that their communities were adequately

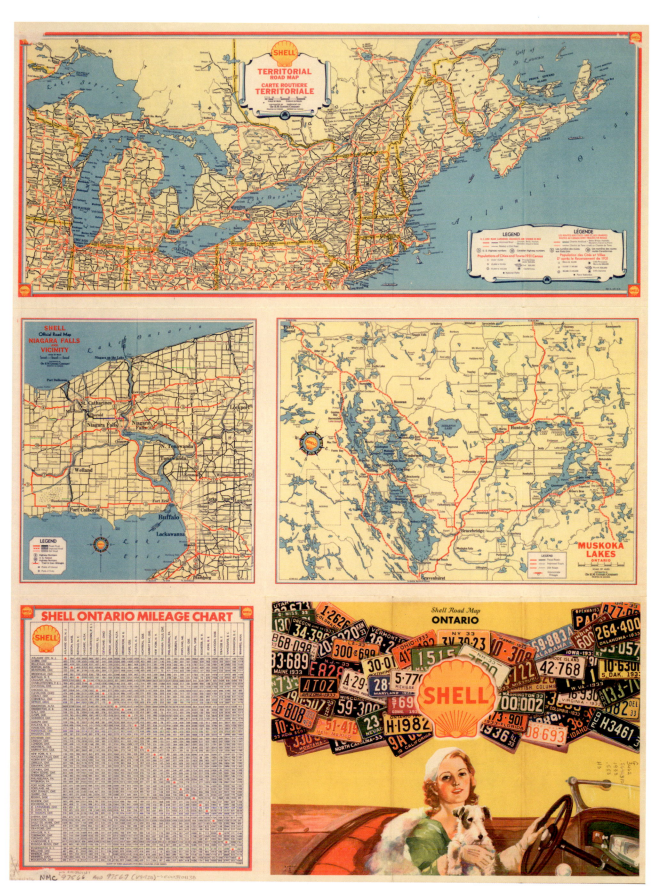

In an era when most women travelled as passengers or as co-pilots, cover illustrations featuring a confident member of the "fairer sex" as driver blossomed in the 1930s. Perhaps the oil companies were trying to suggest that their stations were clean, safe, and friendly destinations for women.

Shell Oil Company, *Shell 1933 official road map, Province of Ontario*, H.M. Gousha Company, 1933, NMC 97567

Confusion reigns for these young cyclists on a tour of the Niagara Peninsula in the pre-automobile years of the late 1890s. Without maps to guide them, such scenes could be expected when "city boys" ventured into farm country. A growing demand for better roads and improved signage, first from cyclists and later, motorists, transformed Canada's rural landscape.

George Schofield, *Cyclists on tour of Niagara region*, ca. 1890s, PA 196531

represented. These self-appointed watch dogs could be extremely passionate in their criticisms. Such was the case when the editor of the *Perth Courier* discovered that the Ontario Motor League had omitted both Perth and Carleton Place from its 1926 directional guide for Highway 15—the new provincial highway connecting Kingston and Ottawa. The editor called the error "a gross injustice" and an "undeserved insult," and went on to demand the recall of all copies of the map "before further harm is done to these two towns in helping to deprive us of tourist traffic." His vendetta had its intended effect. Two months later the League acknowledged its mistake and on December 24 a somewhat smug editor announced to his readers that Perth "will not be ignored in the new road map for the use of tourists."

Like Hamilton's Royal Connaught Hotel a decade and a half earlier, the auto-touring industry was also

beginning to recognize the value of using free road maps in their marketing programs. Imperial Oil Limited was among the first. In 1930 it issued a four-sheet road map of Canada that was hailed by the *Globe* as "one of the outstanding publications this year." Scaled at 1:1,900,800 (almost 1 cm to approximately 20 km), the four sheets together represented an area of over 3.2 million square kilometres and embraced Canada from coast to coast. Published for the company by the Toronto printers Rolph-Clark-Stone Limited, the map was distributed free to motorists at all Imperial Oil filling stations in Canada. Using a tasteful light-green background, it showed all the major border crossings with the United States and colour-coded Canada's highway system into main roads (red) and secondary roads (black). Mileage and highway numbers were included (also in red) and differences in the population of Canada's cities and towns were indicated in typeface sizes (which at the time was a relatively new concept in cartography). The reverse of the map was equally attractive and informative, featuring road plans of principal cities, district maps of summer playgrounds, information on the provincial game laws, and, of course, advertisements for specialty auto products sold by Imperial Oil.

Within two years the Shell Oil Company and the Canadian Tire Corporation followed Imperial's precedent with their own free road maps: Shell issued provincial road maps of Ontario and Quebec and a detailed map of the Montréal region, all printed by the H.M. Gousha Company of Chicago; Canadian Tire published a map of Ontario.

The free road map was now central to the touring industry's marketing of its products and services to the Canadian public. With this transformation, cover art became richer and jazzier, moving away from simple, unillustrated covers to imagery that focused on Canadians' growing love affair with the automobile and motoring adventures. While mapmakers worked hard at accommodating travellers' needs for precise information without comprising the map's clarity and readability, graphic artists laboured equally at romanticizing automobile travel through captivating, colourful covers depicting the great Canadian landscape. Illustrations of a lone automobile, free to explore the countryside in any direction, to stop at will and drink in a rich variety of scenery were favourite themes with publishers.

Much to the delight of the oil industry, Canadians readily took to their maps, which, after the Second World War, oil companies began to turn out in the millions each year. Surprisingly, the road map became synonymous with motoring pleasure and, as the baby boom shifted into overdrive, it was the gas station and its maps to which Canadian families turned for cartographic wisdom. "Make sure every member of the family knows how to read a road map," advised the woman's travel authority in her 1958 column for the *Renfrew Mercury*. "So everyone helps the driver on long trips. Even the children contribute. Each day, someone is the 'navigator,' keeping the map and telling the driver about turns, route numbers, etc. This makes the children better passengers."

When the 1960s put at least one car in every driveway, oil companies found that they no longer needed to seduce Canadians into traveling; they were finding the open road without any encouragement. Road map covers reflected this transformation, changing from messages that stimulated travel to those focused more on attracting customers to their particular band of petroleum products. In such a competitive environment, map covers helped oil companies lure motorists past competing gas pumps. A good road map bred potential customers no matter what their age.

In the late 1970s the free road map went the way of the Studebaker and the Stanley Steamer. The Arab oil embargo and gasoline shortages did away with much of the intense competition that had previously spawned the give-away map. Although road maps are still available at most service stations, they usually cost several dollars and will have been produced by a third party with little interest in promoting motoring pleasure or one brand of gasoline over another. Nevertheless, the road map has left its mark on Canadian society. The once ubiquitous free map helped to produce an entire generation of amateur explorers. Relegated to the back seat as children, they whiled away the hours following the lines, dots, and numbers to imaginary places. ❧

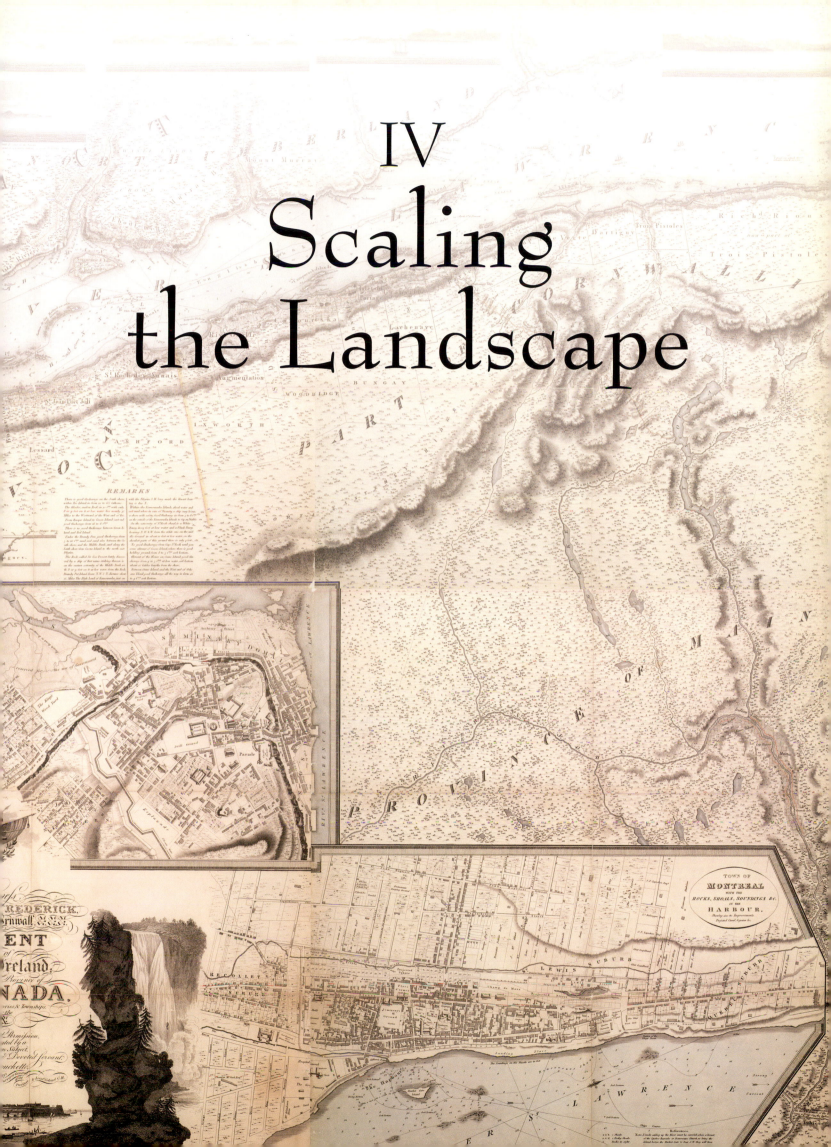

IV
Scaling
the Landscape

CANADA IS IMMENSE, stretching more than 4,600 kilometres from the northern tip of Ellesmere Island to the southern tip of Point Pelee, and 5,047 kilometres from St. John's on the Atlantic coast to Victoria on the Pacific. At 9.9 million square kilometres, it takes up half a continent and could easily accommodate much of Europe. The geographical diversity of the landscape is equally remarkable. Canada's towering mountains, northern tundra, expansive prairies, and majestic lakes are unmatched. Prime Minister Mackenzie King was not too far off the mark when he once commented that, where some countries have too much history, Canada has too much geography.

The immense and diverse landscape has also been a curse. From the first plotting of the country's eastern shores by Johannes Ruysch in 1508, it took mapmakers some three and a half centuries just to complete a basic outline of the coasts. The search for a Northwest Passage, new fur-producing areas, a route for a transcontinental railway, the safe navigation of coastal waters, and untapped mineral resources provided an incentive for much of this mapping. Only after information slowly accumulated on the outline were mapmakers able to turn to more detailed studies of the land's physical features. At first, large-scale mapping was limited to lone settlements and harbours along the seaboards, and was soon followed by the regional surveys of James Murray and Joseph Bouchette. Although each of these projects constituted a milestone in Canada's cartographic history, they did not cover even 1 percent of the total land base that needed to be described.

More expansive detailed surveys were eventually begun in the 1870s and 1880s, when the federal government's Department of the Interior undertook a massive cadastral survey of the Prairies. However, not until the creation of the interdepartmental National Topographic System (NTS) in the 1920s was there an effort to initiate a systematic, detailed mapping of Canada's entire landscape under a unified standard. The horrific battles on the Western Front during the First World War provided the incentive for the NTS program. With the help of aerial photography

As the Surveyor-General of Canada from 1803 to 1814, Joseph Bouchette had this miniature portrait painted during his 1805 visit to London. It was used in the engraved frontispiece of his monumental *Topographical Description of the Province of Lower Canada* (London, 1815). Although the 700-page volume and accompanying map won a gold medal from the Society for the Encouragement of Arts, Manufacturing, and Commerce, it still described less than 1 percent of Canada's total land mass.

John Cox Dillman Engleheart, *Joseph Bouchette*, watercolour on ivory, 1815, C 112049

Noted for its detail, Bouchette's large-scale map identified settlements, farms, roads, fields, vegetation, and important buildings such as mills, churches, blockhouses, and taverns. Changes in topography were indicated through hachures.

Joseph Bouchette, *To His Royal Highness George Augustus Frederick, Prince of Wales, Duke of Cornwall,... this topographical map of the Province of Lower Canada, ... is ... most gratefully dedicated by ... Joseph Bouchette, His Majesty's Surveyor General of the Province & Lieut. Colonel C.M.*, London, J. Walker and Sons, 1815, NMC 21031

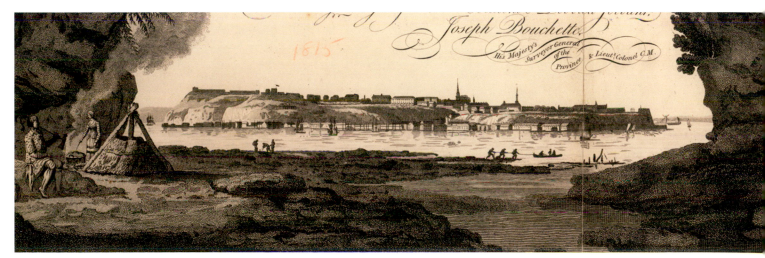

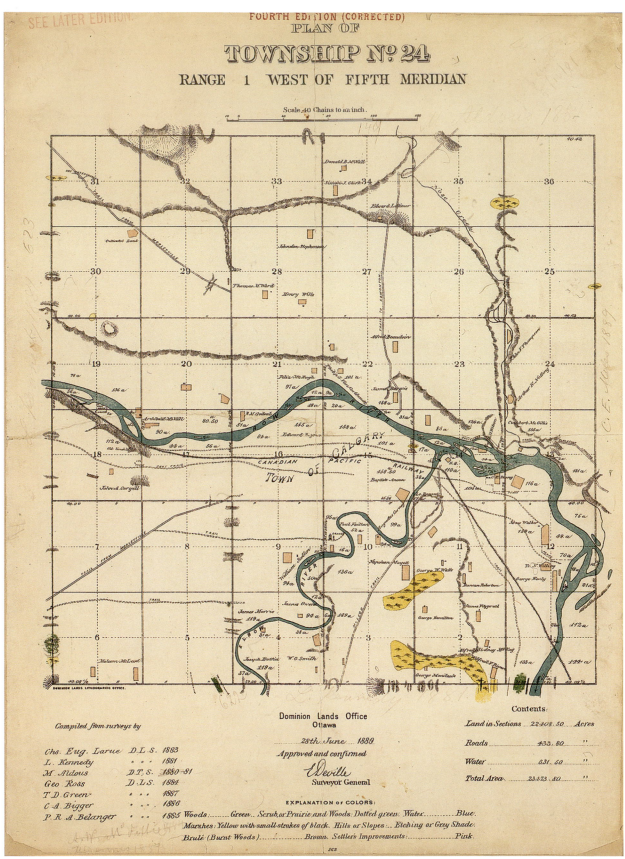

This cadastral plan of Calgary was among the one million large-scale township plans of western Canada produced by the federal government over the 60-year period ending in 1930. The survey prepared the Prairies for settlement by European immigrant homesteaders. Despite the immensity of the project, it covered only about 10 percent of Canada's land mass.

Canada, Dept. of the Interior, *Plan of township 24, range 1, west of the fifth meridian*, Ottawa, Dept. of the Interior, 1889. From *Township plans of the Canadian West*, 1871–1930s, NMC 26423

(and, more recently, satellite imagery), it took almost half a century to implement the plan at the 1:250,000 scale (1 cm to 2.5 km) and another thirty-five years after that to implement it at the more detailed 1:50,000 scale (1 cm to 0.5 km).

This section looks at three large-scale mapping projects. The first considers a cadastral mapping that was undertaken in the counties of eastern Canada in the mid nineteenth century. Cadastral maps, which typically show property boundaries and sometimes the owner's buildings, were originally devised by European governments for taxation purposes. In Canada, however, commercial mapmakers started the first systematic cadastral mapping of the country. These maps had little to do with taxation, and were offered mainly to rural clients through an elaborate marketing scheme that allowed patrons to celebrate their achievements as pioneers.

Two other chapters focus on large-scale topographic mapping by the federal government. Topographic maps have been called the most exacting of all forms of cartography because of the difficulties of depicting the three-dimensional ground surface on a two-dimensional piece of paper. In the nineteenth century, Canada's topographic maps commonly showed relief through hachures, markings that consist of short lines drawn in the direction of the slope (the steeper the slope, the steeper the hachures). Since it is impossible to assign altitudes to hachured slopes, the depiction, while realistic, did not offer the precision that some situations required. For this reason, contour maps are now the topographic map of choice. With this method, relief is represented by a common line, or contour, that interconnects points of equal elevation. A well-drawn contour map will provide accurate elevations for all hills and valleys in a landscape.

In the past, contour mapping necessitated a detailed field survey, which was time-consuming and expensive, especially in hilly terrain. The second chapter examines how Canadian mapmakers overcame these problems in the western mountain ranges by using photography—a technology that proved to be easily adaptable to airplanes and, less than fifty years after aerial photography was introduced to Canada's mapping scene, would provide a complete scaling of the Canadian landscape. The third chapter explores how Canadian technology helped mapmakers on the Western Front provide British and Canadian artillery with the contour maps they needed in order to develop methods of indirect fire on German trenches. ∾

CHAPTER 10

Hawking County Maps

"THIS IS PROGRESS in the right direction," declared the editor of the *Halifax Sun and Advisor* when in September 1865 he finally got a chance to see Ambrose Church's map of Halifax County. Calling it a "splendid map … as near topographical perfection as it is possible to be," the editor could hardly contain his enthusiasm for this "tedious and difficult, and enormously expensive" mapping project. "The whole county has been surveyed and mapped by men specially trained to the business." As the editor saw it, Halifax County was now in the running with many of its American neighbours. "Everywhere one travels in the Northern States," he reminded his readers, "a local map suspended in offices, dwellings, hotels, etc., shows every

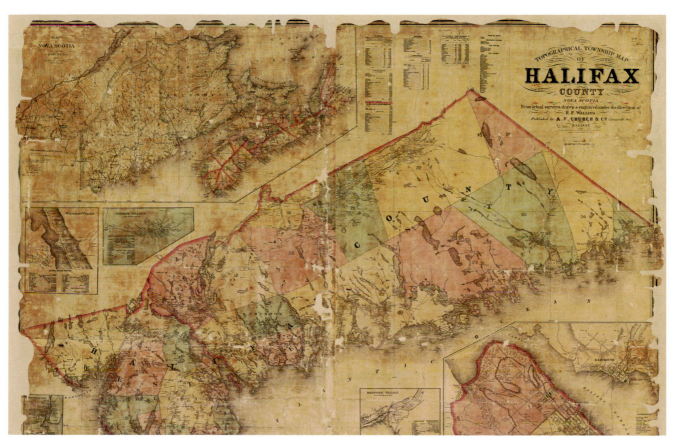

Ambrose Church's 1865 map of Halifax County, Nova Scotia, was widely endorsed by newspaper editors of the day.

Ambrose F. Church, *Topographical township map of Halifax County, Nova Scotia, from actual surveys drawn and engraved under the direction of H.F. Walling*, Halifax, 1865, NMC 111189

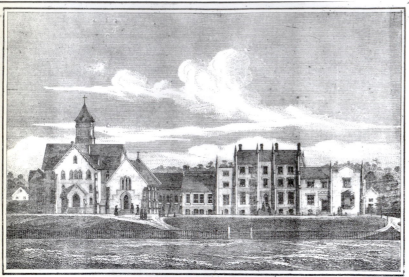

road, dwelling, harbor, wharf, etc., with wonderful
accuracy." And now, because of Church's labour,
Halifax County had a map "in the same style."

Thanks to Church and handful of other mapmakers
from the United States, rural families in eastern Canada
were introduced to a host of new maps and atlases
in the late 1800s, featuring everything from the
county next door to Civil War battlefields. Governments
throughout North America were heavily committed
to national development, and more interested in defining the unknown portions of the western frontier than
the settled regions of the east. Recognizing this imbal-
ance, private publishers like Church capitalized on it
by producing maps and atlases that could be used in
homes, businesses, and schools.

Large wall maps of individual counties were particularly useful to eastern audiences, as they showed
the county's major geographical features—waterways
and hills—in addition to roads, railways, churches, post
offices, schools, mills, and factories. What made them
unique, however, was that they also identified rural
property boundaries and the owners' names. In effect,
county maps were equivalent to a modern telephone
directory. But unlike today's phonebook, they were not
merely utilitarian. Publishers went out of the way to
make the maps as visually pleasing as they were informative, filling them with finely engraved illustrations
of the county's natural wonders, important businesses,
public buildings, and thriving estates. Sometimes the
maps even included short historical sketches about
some of the larger settlements and leading citizens.
Together with their fancy lettering, decorative borders,
and hand-colouring, they conveyed a sense of a community's importance and prosperity.

Almost without exception, the maps were sold on
a subscription basis with one-third or one-quarter of
the basic price of $6 paid in advance and the remain-
der due on delivery, usually a year or two down the
road—or at least after the next harvest. Advance sales
supplied some working capital to finance the expensive
task of engraving the copperplates, and provided publishers with an indication of how many copies had to
be printed. The subscription list also gave the mapmaker a marketing tool. By printing the subscribers'
names in a special panel on the map, the publishers
flattered their patrons' vanity by offering them a chance

to associate themselves with prominent members of the community. For county residents, the maps were not only a geographical reference tool, but also a visual reminder of their community spirit and success as pioneers.

To sell their maps, atlases, and works of art, publishers suggested that their agents prey upon the human need to be recognized and accepted, and the desire to ascend the social ladder. "You should see the most prominent and influential people first, and it is very essential that you should have your best description [of the item you are selling] to give them," one publisher advised its agents. "When every argument has failed, people will sometimes change their mind, and wish a copy, on being shown the names of people who have taken the work." Many potential customers were so mindful of social status that they would go to considerable trouble and expense to add their name to an agent's subscription list so that their neighbours would see it.

Another publisher suggested that, when a sale was made, agents should try to take advantage of the customers' vanity and persuade them to recommend friends who might also be interested in buying the work. "You can expect that [the customer] will do all in his power to urge others to put down their names" in order to uphold the reputation of the published work, "as he would not like to be laughed at for purchasing a worthless one."

Annie Nelles learned first-hand how well this advice worked. Finding it difficult to make any headway, she hit upon the idea of approaching one of the "leaders of society," a Mrs. Rodgers. Nelles offered Rodgers complimentary publications in exchange for room and board. She then proceeded to tell every lady whom she called on that Mrs. Rodgers owned copies of these works. "A furore was created," announced a very confident Nelles as the ladies vied with each other for her own copy. Elated over the success of her method, a smug Nelles concluded that "society … is very much like a flock of sheep; they will stand huddled together … until some one … makes a break in some particular direction … This gregarious disposition of the human race has been of immense service to me."

In the mid 1870s, about a decade after Church had mapped all eighteen counties in Nova Scotia, publishers began to realize that if they printed county maps in an

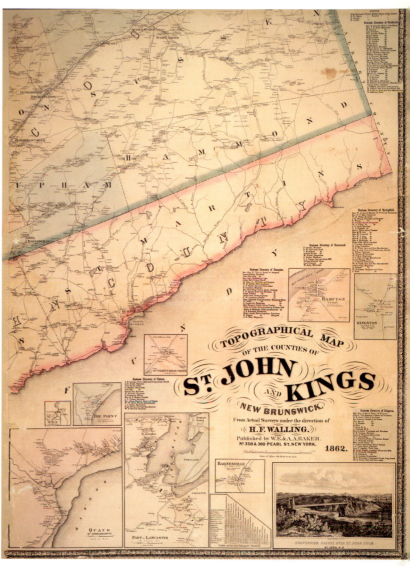

This portion of H.F. Walling's 1862 map of St. John and Kings counties, New Brunswick, featured all the extras publishers liked to include for additional fees: a view of a public landmark, map insets of the towns, and business directories.

Henry Francis Walling, *Topographical map of the Counties of St. John and Kings New Brunswick from actual surveys under the direction of H.F. Walling*, New York, W.E. & A.A. Baker, 1862, NMC 13805

atlas format, they would be able to include a larger number of illustrations and biographical sketches. Since they could charge clients extra to have these items included in the atlas, the profitability of their publishing venture increased substantially. Of course, the atlas format was slightly more expensive to produce and new marketing strategies had to be developed to convince farmers and businessmen to increase the amount of their subscriptions.

Publishers were always on the lookout for sales agents to walk back roads and collect subscriptions for

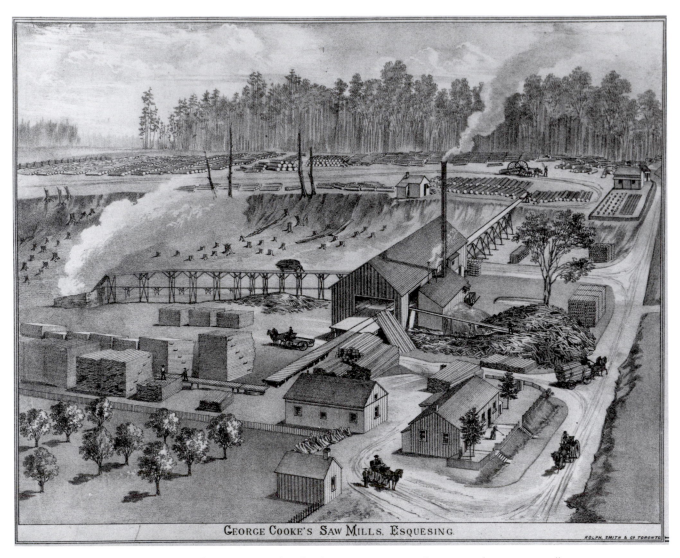

For many 19th-century businessmen such as George Cooke of Halton County, Ontario, the county atlas was an excellent advertising medium. It allowed him to demonstrate his civic pride while boasting of his success as a businessman.

Unknown artist, *George Cooke's saw mills, Esquesing*, Toronto, Walker and Miles, 1877. From *Illustrated historical atlas of the county of Halton, Ont.*, 1877

maps and atlases. More often than not, agents would rub shoulders with others out to hawk dictionaries, encyclopaedias, songbooks, memoirs, and art books. The door-to-door book trade reached its peak in the late nineteenth century, and competition could be fierce as publishers tried to cash in on rural Canada's insatiable desire for "culture." As early as 1871, the *Canadian Bookseller* was reporting that a print run of 3,000 to 5,000 was standard for most works sold through bookstores. However, if the title was marketed door-to-door, print runs could be expected to jump to an impressive 10,000 to 20,000 copies. County maps and atlases fit nicely into this scheme of things: simply by owning them, a family would be seen by its neighbours as well educated. A library nicely stocked in

maps and leather-bound atlases brought its owners a degree of social standing, even if the items sat unread on the shelf. Best of all, they looked impressive in the front parlour!

Recruitment circulars for agents portrayed the door-to-door business as an honourable employment that benefited all "mankind." The accomplished canvasser, boasted one pamphlet, "is necessarily intelligent in mind, courteous in manner, a benefactor to others." In another, the Rev. John Todd announced that he rejoiced "in this carrying the waters of knowledge to the very doors of the people, and almost coaxing the people to drink." He went on to call it "the beginning ... of a great system of creating an enlightened community."

Mapmakers found that one of the most effective promotional gimmicks was the odometer they pushed in front of them as they walked the county's roads measuring distances. Not unlike a wheelbarrow in appearance, it consisted of a small brass instrument with a series of cogwheels. The perimeter of the odometer's wheel, when multiplied by the number of revolutions recorded on a dial, indicated the distance the instrument had travelled. Although less accurate than the compass-and-chain method used by government surveyors, the odometer was economical to operate. One person with a compass and odometer could do the work of a party of three with a compass and chain. Given the scale at which the county maps were published—usually between 1:40,000 and 1:100,000—errors in measuring distances could be kept to less than 2 percent.

The odometer attracted considerable attention. Most farmers had never seen such a device and would willingly stop work to have a good look, usually just enough time for the agent to introduce himself, give his sales pitch, and obtain a subscription to the map or atlas he was preparing. If anyone should doubt the accuracy of his work, he had only to show his sophisticated measuring device.

No matter what the client's social status, the agent's objective was always the same: get the customer to agree to buy the item. Only when the terms of delivery and payment were worked out would the agent mention the more expensive options available. In the case of large county wall maps, options might include varnishing to protect the surface or various types of endpieces that helped in rolling and unrolling the map.

Peddlers of county atlases took a somewhat different approach. Any farmer or businessman placing an order for a county atlas would be visited at a later date by other salesmen asking him to consider including his own family's history to the book—for an extra fee. The salesman would lead the farmer to believe that adding his life story would greatly improve the work's historical value, recording for future generations his

THE ODOMETER.

County map surveyors used odometers to measure distances along roads. The device provided a reasonably accurate measurement, and more important, helped to catch the attention of potential subscribers.

Unknown artist, *The odometer*, Syracuse, W.I. Pattison, 1890. From *How 'tis done, a thorough ventilation of the numerous schemes conducted by wandering canvassers ...*, 1890

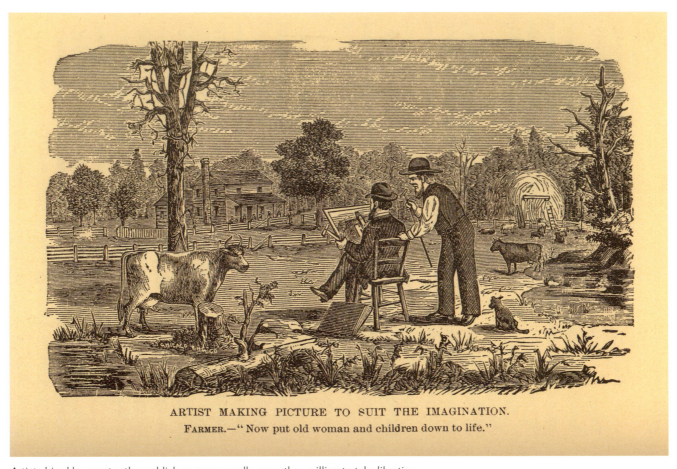

ARTIST MAKING PICTURE TO SUIT THE IMAGINATION.
FARMER.—" Now put old woman and children down to life."

Artists hired by county atlas publishers were usually more than willing to take liberties with their sketches to suit the tastes of their clients.

Unknown artist, *Artist making picture to suit the imagination*, Syracuse, W.I. Pattison, 1890. From *How 'tis done, a thorough ventilation of the numerous schemes conducted by wandering canvassers ...*, 1890

outstanding contribution to the county. A third agent might visit the same farmer to solicit his portrait or a view of his farm or business. An artist was always willing to make any changes or embellishments to the drawing the owner wished, such as adding a pump to the well or replacing dilapidated rails with a neat picket fence. With a bit of coaxing he might even make the barn larger, give the house a coat of paint, and install a lawn. The view is always aesthetically pleasing and the property in an impossible state of tidiness and surrounded with symbols of prosperity: sleek and angular livestock, the latest farm implements, a well-landscaped yard, a hired hand tending the fields, and family members passing the time in leisurely pursuits.

The salesman's slick appeal to the farmer's vanity was difficult to resist, and it was not unusual for some subscribers to a county atlas to end up paying an additional $200 to $300 above the basic $15 cost of the book. If a potential customer was initially approached with these expensive extras, there was a good chance he would refuse to purchase the work. But taken one step at a time, it was considerably easier for a "fast-talking" agent to increase the amount of the farmer's commitment.

The agents certainly were fast-talking. "People expect to hear you talk," advised one publisher's how-to pamphlet for agents, "and you must not disappoint them … keep pouring *hot shot* at them." Publishers suggested agents should try to attract the prospect's undivided attention by cornering him in a fenced yard, behind a stump, or on a plough beam—and then keep talking. Never give up was the agent's unwritten credo. "Four fifths of all the failures in this business, we venture to assert, are attributed to nothing but a lack of that unyielding perseverance," asserted another publisher's pamphlet.

Although door-to-door selling was lucrative for publishers, it did not always turn out the same for agents. "When we employ new men," reflected one

DOOR-TO-DOOR PEDDLING

Who were the door-to-door agents who willingly canvassed rural Canada for weeks at a time and whose only tools were a sales contract and a handy pen? In an 1881 interview, one member of the breed confessed to a reporter for the *Globe* that publishers' agents generally fell into one of two types. "Those who like the work, and have adopted it as a permanent vocation" and those "who take up book canvassing as a last resort or a temporary employment till something else opens." Unfortunately, this last type of agent quickly brought the door-to-door business into disrepute. "They have no liking for the work," admitted the agent to the *Globe*, "and go at it very often in a spirit of desperation—not caring who they offend, or whether they leave a favourable impression or not, so long as they can make a few dollars."

The farmers of Stratford ran into one such agent. He passed himself off as a representative of the Ontario government and carried a questionnaire asking people in the region whether they felt a new survey of the province "with the view of issuing official maps with all the township lines, etc." might be worthwhile. Seeing no objection to this "government proposition," most farmers signed. Imagine their surprise when they received word that the map they had ordered would be delivered by mid summer and would cost $1.90. "The farmers are looking with blood in their eyes for that map agent," reported the *Dutton (Ontario) Advance*, "and not a few of them are indulging in football simply for the benefit of the physical development of the man who will deliver the map."

When the *Winnipeg Daily Times* declared publishers' agents along with mosquitoes, houseflies, and mad dogs the summer's worst pests, there were probably few readers who disagreed. It is no wonder the *Brandon Mail* concluded that "there is probably no class of workers … that has received more abuse, has had more ill-natured flings thrust at it" than the door-to-door agent.

Once door-to-door agents became a regular feature of 19th-century rural life, some property owners went to great lengths to discourage their visits. This type of canvassing was one of the few occupations in the 19th century where a woman received equal compensation to a man for the same amount of work.

Unknown artist, *Warning to agents*, Syracuse, W.I. Pattison, 1890. From *How 'tis done, a thorough ventilation of the numerous schemes conducted by wandering canvassers …*, 1890

SUBSCRIBER'S WIFE.—"Fifteen dollars for that 'ere big book, and too poor to ever get a cook-stove, or fix the house up!"

COMMON SCENE IN DELIVERING ATLASES.

As suggested by this 19th-century engraving, all were fair game for the persistent map agent, whether they could afford the purchase or not. Such encounters eventually gave the trade a bad reputation.

Unknown artist, *Common scene in delivering atlases*, Syracuse, W.I. Pattison, 1890. From *How 'tis done, a thorough ventilation of the numerous schemes conducted by wandering canvassers …*, 1890

A portrait of J.B. Shirtliff—farmer, justice of the peace, and militia captain—adorns a sketch of his 200-hectare estate in the 1881 atlas of Quebec's Hatley Township. Such sketches, and accompanying biographies, remain treasured records of family history.

The residence of J.B. Shirtliff, Esq, Hatley Tp., Quebec, (Massawippi and Ayers Flats in distance).
From Illustrated atlas of the Dominion of Canada containing authentic and complete maps of all the provinces, the North-West Territories and the island of Newfoundland ..., Toronto, 1881

recruiter of agents for a Chicago publisher, "we know in advance that out of every hundred thus employed, five will remain over a month, and one will meet with reasonable success, and that out of every hundred of that successful class, one will develop into a 'star.'" Most agents ended up like Corydon Fuller. The Colton Atlas Company had guaranteed Fuller a handsome profit of $6.50 for every atlas he sold at $15.00 "If I am only well," he wrote in his diary, "I can not fail to make money." In reality, after months of canvassing, his meagre sales barely covered expenses for food, lodging, the upkeep of his horse, and freight charges on shipments from Colton.

County atlases were, at one and the same time, a reflection on the past and a tribute to the families who had persevered and survived. As the introduction to the Perth County atlas stated, the main purpose of the book was "to preserve ... a record of [the county's] early history now existing only in memories of more aged settlers, or in scattered or detached fragments, private memoranda or records which are gradually wasting away." Most of the contributors were the direct descendants of the original pioneers, those who felt that their families' experiences in settling the county were important and should not be forgotten.

Because the books were sold on a subscription basis, the amount of information included on each family depended upon how much space they were willing to purchase, not on their contribution to the community. The result was an uneven distribution: some families were represented by only a few words, others by several pages as well as illustrations. Families who lived outside the dominant society—Aboriginal people, for example—were usually not represented.

Sometimes the biographical sketches were written by family members, but more often than not, they were compiled by someone hired specifically for the purpose. The writer for the Elgin County atlas, for example, was a schoolteacher who had taken on the job for extra income. Despite his professional qualifications, he readily admitted that his sketches "were written for the most part hastily, during evenings after school labour, with little opportunity to gather all the facts, and still less to submit the manuscript[s] after they were written, and as a consequence the large majority of persons never saw the sketches until they appeared before the public."

Because these books were compiled from information provided by the family, they were to a large degree self-censored. The contributors instinctively knew that whatever was recorded would be read by other members of the community and by future generations within their own family. The story they told the writers was probably factual, but it was also slanted to reflect the way the family wanted it remembered. The more painful episodes and sacrifices of forging a new life in the wilderness were either conveniently neglected or overly glorified. The deaths of loved ones, financial problems, and conflicts within the community were seldom, if ever, mentioned.

The historical sketches avoided controversy, especially if there were a possibility that another paying subscriber might be hurt. Only when the community was united in its opposition to a controversial topic originating outside the county were objections voiced. The county atlas of Perth, Ontario, for example, felt perfectly justified in describing the Canada Land Company—the colonization company that organized the settlement of the region—as "a huge monopoly of English capitalists" and "the most unconscionable and unscrupulous ring of 'land-grabbers' which this county … has any knowledge of." Yet it fails

By the early decades of the 20th century, the generations that first settled the counties of eastern Canada were but a memory. As a result, county maps lost their decorative embellishments and became more utilitarian in design and layout.

The Map Specialty Company, *Map of Kent County*, Toronto, 1909, NMC 021915

to mention any of the disagreements community members must have experienced while establishing themselves on the frontier.

County atlases and maps represented a variety of things to different people. For the general reader, they had scientific and educational value because they offered the first real summary of geographical information on the county. They were also considered important for the region's economic development since they provided local businessmen with both a guide to their rural clientele and a useful advertising medium. More often then not, however, they were a chance for both rural and urban residents to flaunt their accomplishments and announce their new prosperity to the world. ❧

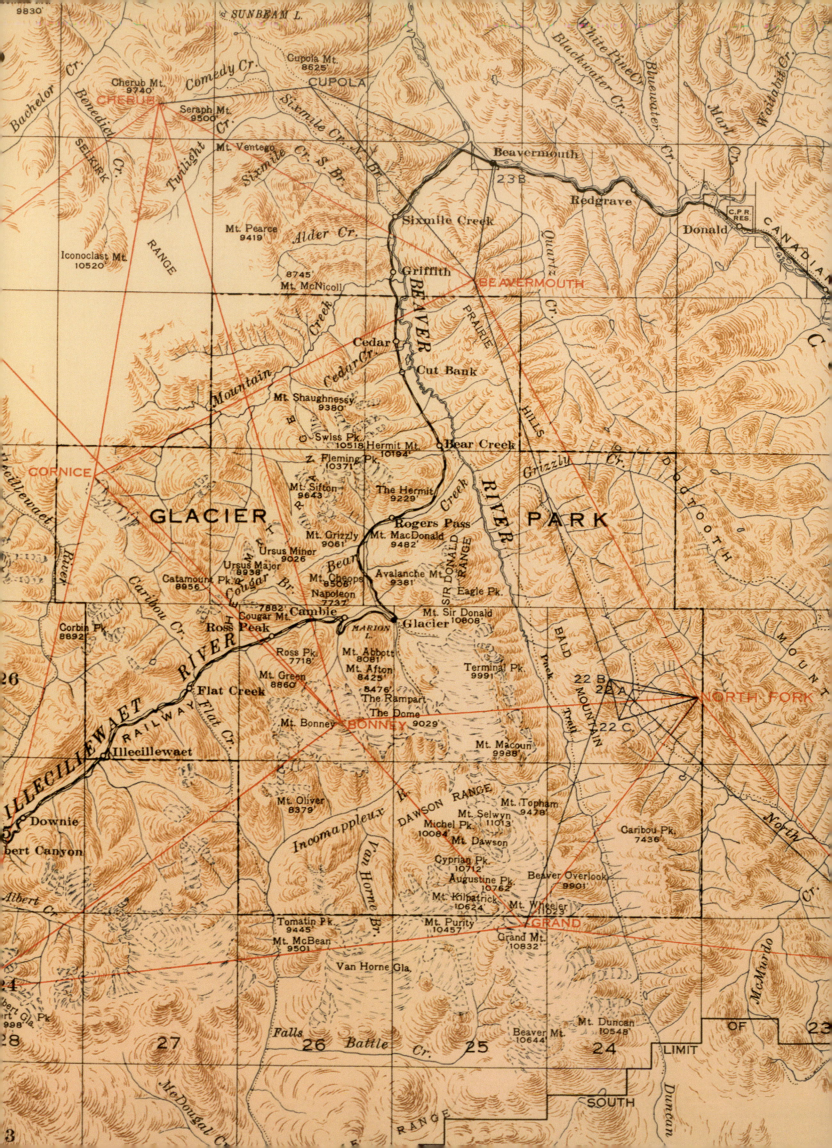

CHAPTER 11

Deville's Topographic Camera

IN THE FALL OF 1925, an inconspicuous wooden box measuring about half a metre square arrived in the basement shipping rooms of the Department of the Interior, the federal agency responsible for monitoring the settlement of western Canada. A red label pasted diagonally across one side of the box simply read: "Glass. Handle Carefully." Although the railway manifest provided no other clues about the contents, experienced clerks knew that the box contained several hundred exposed photographic glass negatives—an entire season's worth of surveys from one of the most rugged and inhospitable mountain systems in the world.

The negatives had been sent from western Canada and were to be used in making detailed maps of the Rocky Mountains. When similar boxes first arrived in 1886, the use of photography in mapmaking was revolutionary. But after thirty-five years of experiments and application by Canadian cartographers, the procedure had become so common that Canada now had an international reputation as a leader in what was widely called "photo-topography." The shipping manifest estimated the replacement value of the glass negatives at $5,000, but to the mapmakers in the building they were invaluable, as without them it would be impossible to make any maps of Canada's western cordillera: the Coastal Mountains, the Interior Mountains, and the Rocky Mountains of British Columbia and Alberta.

From the moment western Canada entered Confederation—beginning with Manitoba in 1870 and British Columbia in 1871—the federal government envisioned an orderly settlement of the region. Homesteads had to be surveyed, a transportation system built, and a system of law enforcement firmly entrenched before mass emigration from Europe would be allowed to proceed. Dominion Land Surveyors divided the western Prairies into a gigantic checkerboard of more than one and a quarter million homesteads in what would be the world's largest land survey grid laid down under a single integrated system. It eventually covered over 80 million hectares—almost one-tenth of Canada's entire land mass. Under federal regulations, every homestead had to be marked first on the ground and then on a map before the land could be opened to settlers.

The prairie landscape presented few problems for surveyors. Open vistas enabled unobstructed sightings and infrequent variations in the near-uniform landscape rarely required mapping. But the situation changed drastically once survey crews reached the foothills of the Rockies in the mid 1880s. The crews were suddenly confronted by a land mass stretching all the way to the Pacific rim, which contained a seemingly infinite number of peaks and valleys in what would prove to be one of the most topographically varied landscapes in North America.

The methods that had been used so successfully on the Prairies—the centuries-old plane-table surveys where distances were measured using a surveyor's chain—were tried but proved ineffective on the steep mountain slopes. Rapid surveys based on triangulations

Édouard Gaston Deville first introduced cameras to land surveyors in 1886 and is now widely regarded as Canada's father of photogrammetry—the science of making measurements and scaled drawings from photographs. Photography is still the basis from which most maps—especially of inaccessible areas—are made today. The medium is also heavily used in ecological studies, for the most part in forestry and wildlife research.

William James Topley, *Edouard Deville, Surveyor-General of Canada*, 1914, PA 043041

was appointed the Quebec government's Inspector of Surveys, the province's top survey position.

Deville had more ambitious plans, however. The same year as his appointment, he wrote and passed examinations in high-level survey theory, mathematics, mapping, and astronomy, and was awarded his certificate as a Dominion Topographic Surveyor (DTS), a coveted title only three other surveyors before Deville had received. Three years later he joined the Department of the Interior in Ottawa to work on the Prairie homestead surveys, and in 1881 the Minister appointed him inspector of federal surveys. Under his tutelage the western surveys expanded exponentially to keep up with the rapid progress being made in railway construction. In 1883, for example, the Department of the Interior contracted 1,200 men (and nearly as many horses) to work on 119 survey crews. Working from sunup to sundown, the crews surveyed some 11 million hectares—a feat that stands unrivalled to this day in North America. In total the crews ran 130,000 kilometres of survey lines and subdivided 1,221 townships. In 1885—the year the Canadian Pacific Railway was completed—Deville was promoted to Surveyor General of Canada, a post that gave him complete authority over the federal government's mapping of western Canada.

Deville thought that photography might offer a solution to the many problems faced by Canadian surveyors in the western mountains. While a student at the French Naval College at Brest, Deville heard of the experimental use of photography in the mapping of Paris by Aimé Laussedat, a French army engineer. Laussedat's *métrophotographie* had been recognized by the prestigious French Academy of Science as early as 1859, and no doubt was hotly debated among Deville's classmates. By the time Laussedat's map of Paris was completed and presented at the Paris exhibition in 1876, he was already well known among European cartographers.

Although Deville greatly admired Laussedat's work—at one point he referred to it as "so complete that little has been added to it since"—he was also a practical man. Deville recognized that the science would require extensive reconsideration if it were to be applied to the Canadian situation. The unmanned balloons and kites from which Laussedat suspended

and field sketches were also attempted but were equally useless—there was too much detail for a surveyor to capture over a reasonable period of time. With their options effectively eliminated, the surveyors faced a depressing future. If the surveys did not proceed, the development of a large portion of British Columbia and Alberta would have to be delayed. And if development were delayed, it would surely mean economic suicide for the fledgling western resource industry.

Fortunately, the agency responsible for all federal surveys in the west, the Department of the Interior, had recently brought on staff Édouard Gaston Deville. Originally a native of France, Deville had just retired from a career on French hydrographic ships in the South Pacific when he arrived on Canada's doorstep in 1874. Almost immediately he went to work as a surveyor and astronomer in Quebec, and in three years

A survey camera and 12 glass negatives were carried into the mountains in a hard leather case that weighed about 9 kilograms in total. An entire season's work could amount to several hundred negatives. All the plates would be packed carefully and shipped to Ottawa, where they would be turned into topographic maps. Many of these original plates from the first photo-topographic surveys are now in the holdings of Library and Archives Canada.

Canada, Dept. of Mines and Technical Surveys, *Construction and case of topographic survey camera*, ca. 1920, PA 023202

his camera over Paris could not be used in Canada's isolated and rugged western mountains. The unpredictable winds in particular would play havoc with the balloons and create an unstable platform for the cameras and their slow shutter speeds.

Wasting no time, Deville started his own photographic experiments during the next field season. He first designed a sturdy but light field camera that could be carried for long distances and recalibrated once the surveyor arrived at his station. The camera consisted of a rectangular metal case open at one end. The opening contained a series of diaphragms that kept out unwanted light, but permitted the insertion of a 10-by-15-centimetre glass negative. The other end of the case accepted a mahogany box that held an English-made shutterless lens.

Deville gave the first camera to James McArthur, one of the department's veteran surveyors. McArthur, as it turned out, was a good choice. Often described as

a "quiet and unassuming man" by his peers, he had probably climbed more mountains in the western cordillera than any other person. Deville's instructions were straightforward: with his one assistant, McArthur was to take panoramic views from the mountain peaks in Rocky Mountains Park—a new park the federal government had established to protect the Banff hot springs.

Unlike Laussedat, who worked primarily from vertical photographs, Deville's method called for oblique (bird's-eye) views of the landscape. With the camera set on a mountain peak and its low-distortion lens carefully levelled and trained on the horizon, the surveyor would take panoramas of the nearby peaks. The precise orientation of each of these views relative to the survey station would be measured using a transit or theodolite (optical devices for measuring horizontal and vertical angles). The process would be repeated from another mountain peak so that each feature

Topographic Lands Surveyor James J. McArthur and his assistant carry their photographic equipment to a mountain peak in Rocky Mountains Park.

Canada, Dept. of Mines and Technical Surveys, *Topographer and assistant, W.S. Drewry and J.J. McArthur ...*, 1887, PA 023141

would be photographed and triangulated from at least two different camera angles. Once back in the office, the surveyor would plot the camera station on paper, orient each of the photographs relative to the angles recorded by the transit, and draw lines out from the station to each significant feature noted in the photograph. The point at which these lines intersected from two or more stations would be the true position of the feature relative to the camera stations.

Although considered experimental, McArthur's initial surveys fully met Deville's expectations. "Instead of the rough and imperfect sketches which such explorations generally furnish," boasted Deville to the Minister of the Interior, "we will have, without extra cost and with but little extra office work, complete maps of the country, which if made with the usual methods would absorb very large sums of money."

McArthur's surveys were allowed to continue for another six years, and in the end they produced a twenty-one-sheet topographical map that covered a 3,400-square-kilometre section of the Rockies—the entire national park, plus a large area outside the boundaries. Scaled at 1:40,000 (1 cm to 0.4 km) and featuring a contour interval of 100 feet (30 m), the map was Canada's first detailed survey of a mountainous region and was widely hailed a success by professional cartographers across the country.

Photo-topography quickly established itself on the Canadian scene because it eliminated one of the major impediments to surveying in the western mountains: the short survey season. In higher elevations the survey season was typically limited to a three-month period. Within this narrow window, unstable weather often meant that at best only a few hours a day could be counted on for observations. "We often reached the [mountain] top ... and remained in mist all day hoping it might clear; only to be obliged to descend without having accomplished anything," complained McArthur after one particularly long stretch of bad weather. When the weather did break, they might have only a few minutes to take photographs before the clouds obscured the view again. "It often took us from three to eight hours to complete our observations," explained McArthur, "and many occasions, after shivering with cold all day on the summits, we had to return to camp and there await another opportunity."

During the 1892 field season, for example, McArthur spent 111 days in the mountains. Out of that total, he lost 48 days to smoke, rain, fog, and snow, leaving him with only 63 actual working days. But he was still able to survey an impressive 1,230 square kilometres. In the 1894 field season, the weather was more cooperative. "I made 23 ascents and exposed 275 plates," wrote McArthur in his report to Deville. "The combined areas covered during this season by the seven [survey] parties equalled nine thousand square miles [23,000 sq. km]."

Like his mentor Laussedat, Deville chose a world's fair—the Chicago Columbian Exposition of 1893—in which to introduce his new mapping technique to the

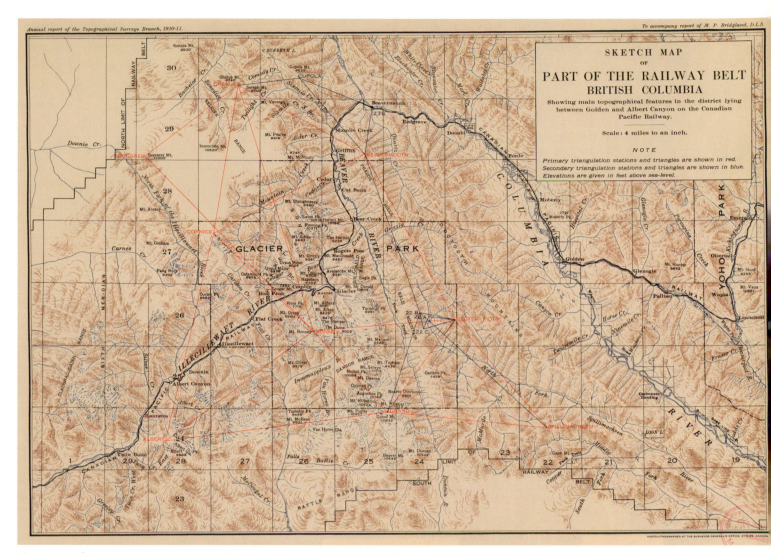

The system of primary (red) and secondary (blue) triangulations established by Morrison Bridgland during his 1910–11 photo-topographic survey of a section of the Railway Belt in southern British Columbia is superimposed on this sketch map of the region.

Morrison Parsons Bridgland, *Sketch map of part of the Railway Belt British Columbia showing main topographical features in the district lying between Golden and Alert Canyon on the Canadian Pacific Railway*, Ottawa, Topographical Surveys Branch, 1910–11, NMC 5850

world. A twelve-sheet topographic map that covered a 1,800-square-kilometre section of the Canadian Rockies was specially made for the exposition. Each sheet was compiled from an average of sixteen camera stations, which provided the mapmakers with up to 120 panoramic views. The cost of the map worked out to an incredible $2.80 per square kilometre. "This extremely low figure," reported Deville, could be attributed to the much shorter time the topographer "took in making measurements. These are left for the winter months, to be attended to in the office." Deville later estimated that work requiring eight days in the field using a plane table could be carried out by means of photo-topography with only two days of fieldwork and three days of office

work. This meant photo-topography was only about one-third the cost of conventional surveys.

Interest in the map prompted Deville to compile a small pamphlet on the methodology, and two years later, a comprehensive text book (which was circulating among professional surveyors nine years before Laussedat published his definitive work on the subject in Europe). Both these publications did much to draw the world's attention to what professional surveyors would later nickname "the Canadian oblique" method of surveying.

By the early 1900s, Deville had attained something of a celebrity status. In 1905 the University of Toronto paid tribute to his "service to the cause of science" by

Deville's Topographic Camera   157

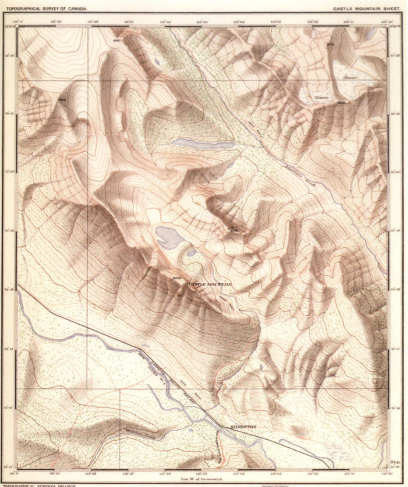

The Castle Mountain sheet from the 21-sheet Rocky Mountains photo-topographic map that was published in the early 1890s. Compiled by James J. McArthur, the map was Canada's first attempt at a detailed description of the topography of the Rocky Mountains.

Canada, Topographical Surveys Branch, *Castle Mountain sheet*, Ottawa, Topographical Surveys Branch, 1890. From *Topographical Survey of the Rocky Mountains*, ca. 1889–92, NMC 9877

conferring upon him an honorary Doctor of Laws; in 1916 a royal warrant appointed him a Companion of the Imperial Service Order in recognition of "the sterling services he has rendered to Canada and to the Empire by his researches." In 1922 he was nominated to represent Canada at the meetings of the International Union of Geodesy and Geophysics held in Rome. His assistance and advice were widely sought by a variety of foreign governments that had initiated their own mountain surveys; however, none of their projects would come close to the amount of mapping Canada had undertaken. The Survey of India purchased a copy of Deville's camera and in 1921 it was used by a British expedition to map some 1,550 square kilometres in the immediate vicinity of Mount Everest. The resulting map was instrumental in a British-led assault on the summit the following year.

Surprisingly, American surveyors showed little interest in photo-topography even though they shared much of the same topography with Canada. With the exception of some surveys done in the early 1900s in connection with the Alaska-British Columbia boundary, they continued with plane-table surveys. One of the few American articles on the use of cameras in surveying, which was published more than thirty years after Canada had embraced the technology, was still calling the technique "experimental" and "in need of further work."

At the outbreak of the First World War, Deville could point to almost a dozen survey parties systematically working their cameras through the western cordillera and to at least a dozen other specialized projects that had been completed—international and interprovincial boundary surveys, engineering studies for irrigation projects, and surveys in connection with resource development among them. As well, in addition to his own department, he could list several other Canadian agencies that had fully adopted the technology: the Geological Survey of Canada, the Geodetic Survey Branch, the International Boundary Commission, and the British Columbia Department of Lands and Works.

Deville died in Ottawa in September 1924, but before his passing Canadian surveyors had used photo-topography to map an incredible 135,000 square kilometres—an area roughly the equivalent to half the size of the United Kingdom. By comparison, camera surveys in the United States had contributed only 32,000 square kilometres of new mapping.

When reliable fixed-wing aircraft came into mass production after the First World War, Canadian surveyors were quick to realize that this means could be used to acquire aerial perspectives similar to those that had been collected from mountain peaks, the main difference being that the airplane was more versatile. It allowed surveyors to move away from the western mountain ranges, and for the first time, acquire oblique aerial views of other parts of the country.

The new technology was invaluable in areas where differences in elevation were minimal, such as the

SURVEYING THE ROCKIES

The Rocky Mountains were Arthur Wheeler's passion. "I do not know how it is with others," he confessed to a meeting of the British Columbia Land Surveyors Association in 1929, "but standing on the summit of a high peak on a bright day with all this glory of intense colour around, I feel it in my inside." As a topographer for the Department of the Interior, and later as a Commissioner on the Alberta-British Columbia boundary survey, Wheeler had many opportunities to experience what he called "the joy of climbing and exploring in these fascinating surroundings." Whether he was dodging avalanches, navigating icy slopes, skirting falling rocks, or crossing glacial torrents, Wheeler thrived on the hazards that the mountains threw at him. On more than one occasion he chuckled when younger men on his crews found the work too daunting and abandoned the survey camp.

Perhaps more than anything, it was the unpredictable mountain weather that set Wheeler's heart pounding with fear. "What we dread most is an honest-to-goodness electric storm," he admitted to the same 1929 meeting, "with lightning bolts knocking up rocks around you and the thunder like cannon discharge close beside you." For Wheeler, a thunderstorm was quite different on top of a mountain then at ground level. "When they pass over," he recounted, "the nearby rocks begin to hum and ice-axes to buzz. If you remove your hat your hair stands on end."

One notable storm caught Wheeler and his crew on Goulds Dome in the High Rock Range. As he recalled, "It was a fine, bright day and we were busy at work ... Suddenly the thunder began to growl in the distance and lightning flashed. The storm quickly travelled toward us and soon enveloped the mountain ... A thunderbolt struck some forty feet from us and knocked up rock and dust ... Another flash, when Cameron sprang up and crying out, "Oh Lord!" clapped his hands to his behind. We all knew what had happened ... We wrapped our coats over our heads and waited for the end ... We thought we should be killed as each flash sent an electric shock through us. Gradually the storm passed and the sun shone bright and clear."

A surveyor adjusts his topographic camera during field trials in the Rocky Mountains. To the left is the theodolite that was used to calculate the bearing (or angle) to other mountain tops.

The Deville camera in operation, 1927. From "Topographical surveying in British Columbia for the National Topographic Map Series," 1927, report found in RG 88, vol. 471

A surveyor's camp in the Rocky Mountains, c. 1920s.

Surveyors camp, 1927. From "Topographical surveying in British Columbia for the National Topographic Map Series," 1927, report found in RG 88, vol. 471

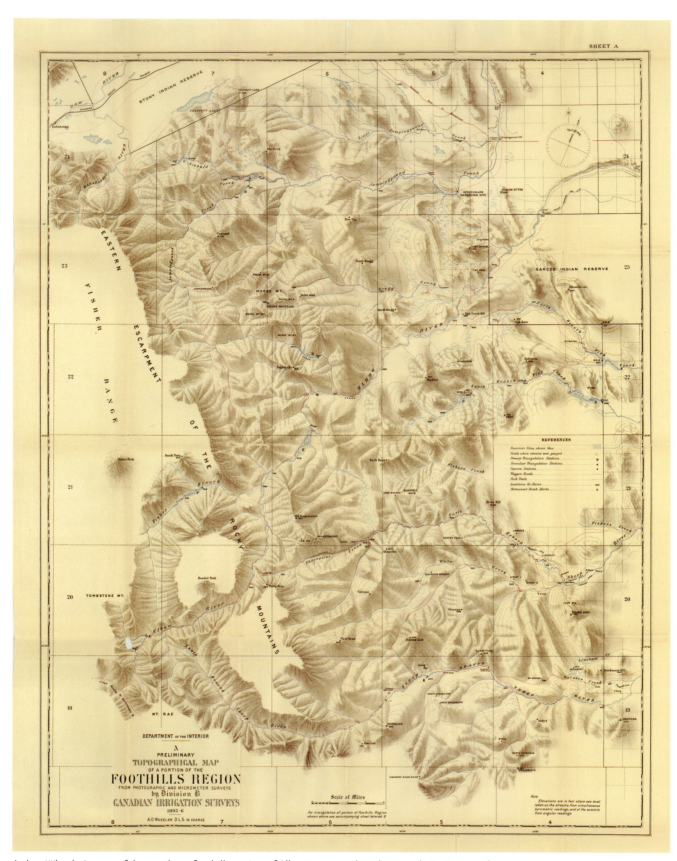

Arthur Wheeler's map of the southern foothills region of Alberta was produced using photo-topography.
The map provided much-needed information for engineers to calculate stream flows and containment ponds
for the irrigation projects that would feed water to the drier grasslands to the east.

Canada, Dept. of the Interior, *A preliminary topographical map of a portion of the foothills region*, Ottawa,
Dept. of the Interior, 1895–96, NMC 23975

The aerial survey crew of an RCAF *Vickers Viking IV* at Victoria Beach, Manitoba. Deville's experiments provided the background technology that enabled cameras to be mounted in aircraft after the Great War.

Canada, Dept. of National Defence, *Crew of Vickers Viking IV flying boat ...*, 1924, C 025910

Canadian Shield and the boreal forest. The hundreds of small lakes that dot these regions were more easily plotted from the air than from the ground, and they functioned as airfields from which surveyors could operate their "flying boats." Despite their versatility, aircraft did not immediately replace Deville's survey camera. On the contrary, the two systems complemented each other for at least three decades. Deville's camera had proved itself very effective in measuring differences in elevation, but was not suitable for providing details of flat areas. Aerial photography, on the other hand, overcame difficulties with flat terrain, but required Deville's camera to establish elevations, especially in mountainous areas.

The two systems were used throughout the western cordillera until the early 1950s, when the miracles of the air age were finally developed to the point that aerial photography could stand on its own as a precise and economical surveying platform for the large-scale mapping of all aspects of Canada's varied landscape. ∾

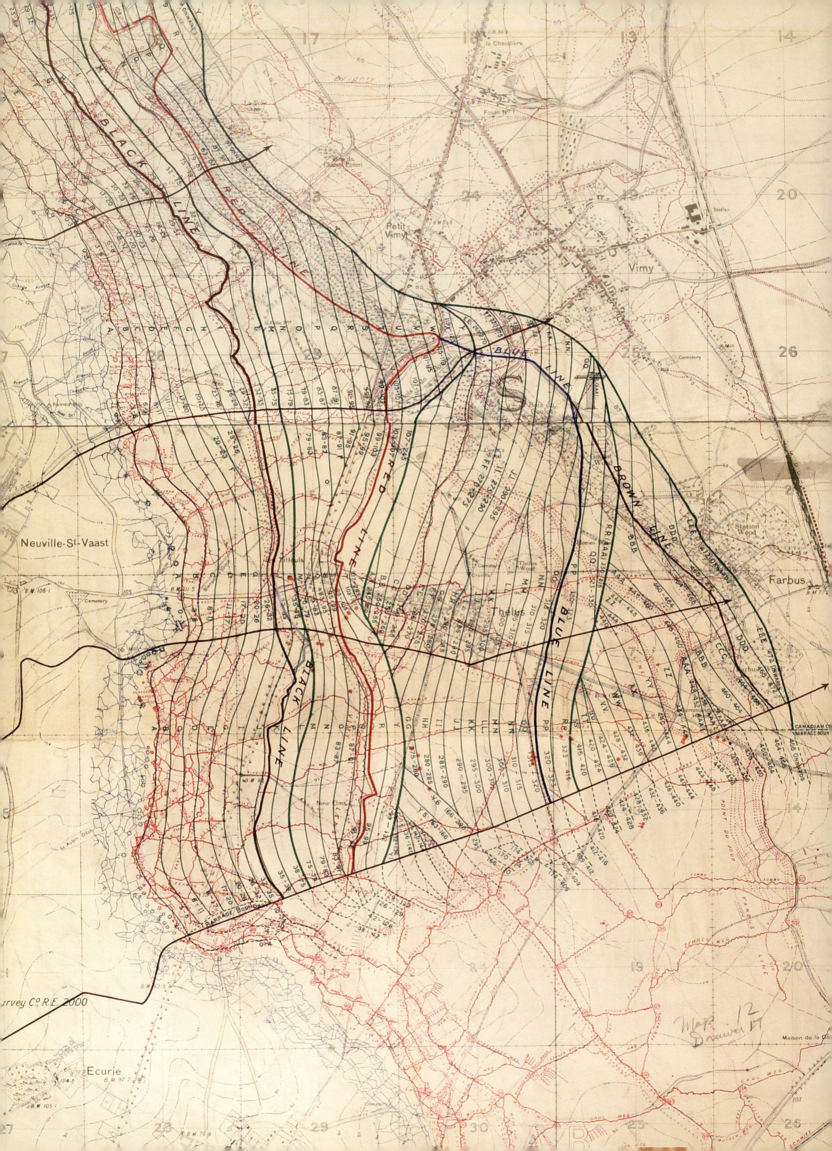

CHAPTER 12

Weapons of the Great War

Lieut. Col. Andrew McNaughton understood only too well the value of accurate topographic maps. As commander of Canada's counter-battery operations at Vimy, he spent much of his time at his drafting table pouring over maps of the battlefield. "What a coldblooded thing war now is," he wrote his wife just a few days before the Canadian assault on the German-held ridge. "We set and plot destruction with pencil on paper, then issue orders and make preparations. More of it tomorrow and still more the day after." In the end, McNaughton's planning proved to be worth the effort. Canada's Easter assault—the first major victory in the lengthy deadlock that had brought the First World War to a stalemate—gained more ground, seized more guns, and captured more prisoners than any previous Allied offensive. Their assault

While waiting for their own artillery and ammunition to be brought forward over the shell-torn battlefield, Canadian gunners use a topographic map to redirect the fire of a captured German gun on Vimy Ridge. The Canadian advance on the Vimy had proved so successful that they moved beyond the effective reach of their own artillery.

Canada, Dept. of National Defence, *An officer directing the fire of a German gun*, 1917, PA 001138

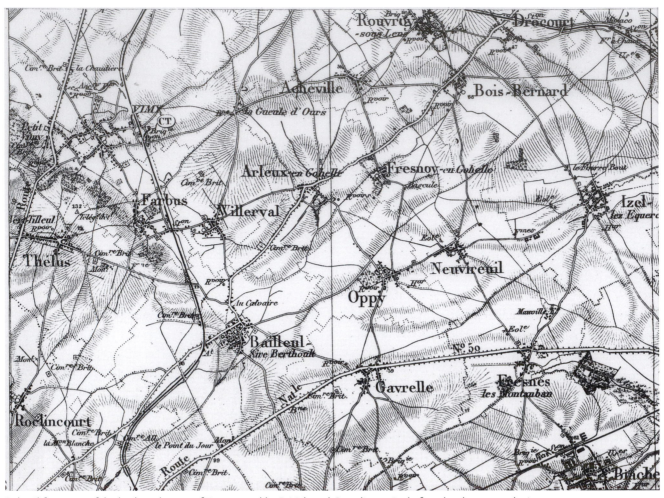

A detail from one of the hachured maps of France used by British and Canadian units before they began producing their own topographic maps. Although they offer a picturesque view of the landscape, hachured maps were of little value to the artillery because they provided no elevations.

Britain, British Expeditionary Force, *Sheet 7, Arras*, Ordnance Survey, 1912. From *France 1:80,000*, GSGS 2526, 1911–12, NMC 114212

swept the firmly entrenched Germans from the 60-metre-high, muddy scarp using 12,000 fewer troops than what the British and French had already lost on the same battlefield in earlier attempts.

Military historians universally agree that the secret of the Vimy success lay in Canada's ability to develop a whole new strategy for employing its artillery. "This was not [artillery] fire preceding movement, as in the early days of the Somme," observed the historian John Swettenham, "but fire combined with movement." Indeed, much of McNaughton's planning for Vimy had focused on accuracy in mapping targets, and on accuracy in shooting once the targets had been located.

The attack on Vimy represented a substantial change in military strategy from the earlier years of the war. When the British Expeditionary Force first embarked for France in August 1914, its leaders were expecting a war of movement. Relying on their Boer War experience, artillery commanders thought they would be fighting in the open where they could observe shell bursts and adjust their fire accordingly. But with the battles of Flanders in the fall of 1914 and the complete entrenchment of both armies along the Western Front, British gunners found themselves in desperate need of help directing their fire—traditional methods of observation were now totally useless.

Fortunately, Maj. H.J.L. Winterbotham, a seasoned surveyor with Britain's 1st Ranging and Survey Section, had recently arrived in France. Winterbotham took a personal interest in the artillery's problem and, after studying the situation, was able to show the gunners that, once their positions were surveyed and their targets plotted, they could easily determine where to shoot by calculating the range and bearing from maps.

For many gunners these new techniques were considered ungentlemanly. Indeed, they were initially so disdainful of Winterbotham's ideas that they nicknamed him "the Astrologer." The infantry were even less appreciative. One of the first survey parties Winterbotham led into the field spent the night in detention, having been arrested for spying by an ammunition column that had become suspicious of their strange-looking instruments.

In order to help the gunners develop methods of indirect fire, British cartographers tried to meet the artillery's need for maps by simply enlarging the French topographic sheets they already had in hand. However, they soon discovered that these items were not suitable for map shooting. Many of the maps had been surveyed almost a century earlier, and some features plotted to an accuracy of only 180 metres. As well, all the maps used hachures—short lines drawn in the direction of the slope—to show ground relief. Although hachures provide a picturesque rendition of the earth's topography, they do not show actual elevations. When shooting from a map, it is absolutely crucial that gunners know the elevation of their target: if it lies on ground substantially higher or lower than their gun, the shot will not fall true unless proper allowances are made for the difference.

Within the matter of only a few months, it had become clear to Allied headquarters that a completely new, large-scale topographic map of the Western Front was needed, and needed immediately. The new map would require a re-survey of more than 31,000 square kilometres of the French countryside, an incredible assignment made all the more astounding by the fact that a large part of the job included German-held ground.

Much to their credit, the Allies began to investigate the possibility of using aerial photographs to help them in this Herculean task. Many Allied pilots were already taking their own cameras on reconnaissance flights and using photographs as supplements to their oral and written reports. They found that oblique photographs, taken from cramped and drafty cockpits, revealed much of what they could not see while fighting the numbing cold and enemy aircraft. Although the quality of the pictures was initially poor, British mapmakers quickly realized that these views of the German trenches were not all that different from the

photographs used by Canadian surveyors in mapping the Rocky Mountains.

Some thirty years previously, Canadian surveyors discovered that cameras could be made into efficient mapping instruments (chapter 11). Largely through the efforts of the Surveyor General, Édouard Gaston Deville, they had managed to develop a camera, marked with a special grid, which could be used in mapping mountain peaks and slopes. The system enabled surveyors to take advantage of the high western peaks in order to obtain a vast panorama of the landscape, a perspective almost the equivalent to that of a low-flying aircraft.

The savings over traditional surveying methods were significant enough that, by the turn of the century, photo-topography was being taught to Canada's

Trained as an engineer at McGill University, Andrew G.L. McNaughton led the 4th Battery to France in 1914. His scientific approach to gunnery eventually won him the command of the Canadian Corps artillery. After the war he continued his military service, rising to chief of the general staff (1929–35) and president of the National Research Council (1935–39). With the outbreak of the Second World War, he became commander of the 1st Canadian Infantry Division.

Pittaway, *Brig.-Gen. A.G.L. McNaughton, Commander, Canadian Corps Heavy Artillery*, 1918, PA 034150

Kite balloons, such as this one under preparation for an ascent, were instrumental in providing the Allies with intelligence information and in directing artillery fire on German targets. The slow moving, gas-filled blimps with their hanging wicker baskets were easy targets for German gunfire; consequently, unlike airplane pilots, balloonists were issued parachutes should they need to make a quick escape.

Canada, Dept. of National Defence, *Pilot and observer in basket of kite balloon*, 1916, PA 000677

military cartographers at the Royal Military College in Kingston, and was part of the regular examination requirements for a Dominion Land Surveyor. The Canadian government had used the method not only to map a large portion of the Rockies, but also some of the Alaska boundary. Because of these initiatives, the mathematical formulae to account for the distortion in oblique photographs were well in place, as were some of the lenses and filters that cut through smoke and haze. Building on this success, the Canadian government had even sponsored a special exhibit on photo mapping at the 1893 Chicago World's Fair.

As a result of these efforts, photo-topography was well known in mapmaking circles by the outbreak of the First World War, even though most European governments had no programs in place. If cameras could be adapted to mountainous conditions, reasoned British cartographers, then surely they could also be adapted to the airplane.

Even the Canadians themselves felt that the only barrier preventing their continued progress in the use of photography was the inability to realize an overview of the countryside beyond the mountain tops. As Major Brock, a former Dominion Lands Surveyor, once recounted: "I remember very distinctly the first time I saw a notice in a paper of an aeroplane making a prolonged successful flight. With my topographer we talked the thing over—we were up in the mountains making a topographic map—and we wondered how long it would be before the Government of Canada could be induced to get a few aeroplanes for topographic work."

The war raging in Europe provided the incentive to send cameras aloft. Although it would be at least another two years before they had a system capable of taking snapshots from the underside of an aircraft, pilots began providing the photographs cartographers needed by the early summer of 1915. But aerial photography was still dangerous work—each reconnaissance plane required the covering fire of five fighters. Canadian war ace Billy Bishop flew escort duty on several occasions. Other Canadian pilots, such as W.G. Baker, R.H. ("Red") Mulock, A.L. MacLaren, and A.G. Goulding, received citations for heroism while flying photo reconnaissance under heavy German fire. Despite the hazards, however, aerial photography proved to be worth the risks. When all the variables were right, the technology could reduce the error factor in mapping to less than 18 metres.

The new topographic maps produced for the Western Front were printed at three different scales: 1:10,000 (or 1 cm to 0.1 km), 1:20,000, and 1:40,000. For the most part, all three maps showed the same basic information, being drafted first at 1:20,000, then either enlarged or reduced to one of the other two scales. Intelligence information on the German lines was added to all three scales as overprints, a clever move on the part of the mapmakers as it enabled them to update their intelligence to reflect the quickly changing situation on the battlefield without going to the trouble of reproducing

Aerial photographs of the trenches on Vimy Ridge aided military survey units in drafting up-to-date detailed maps of German defences.

Canada, Dept. of National Defence, *Trenches on Vimy Ridge from a kite balloon*, 1917, PA 002366

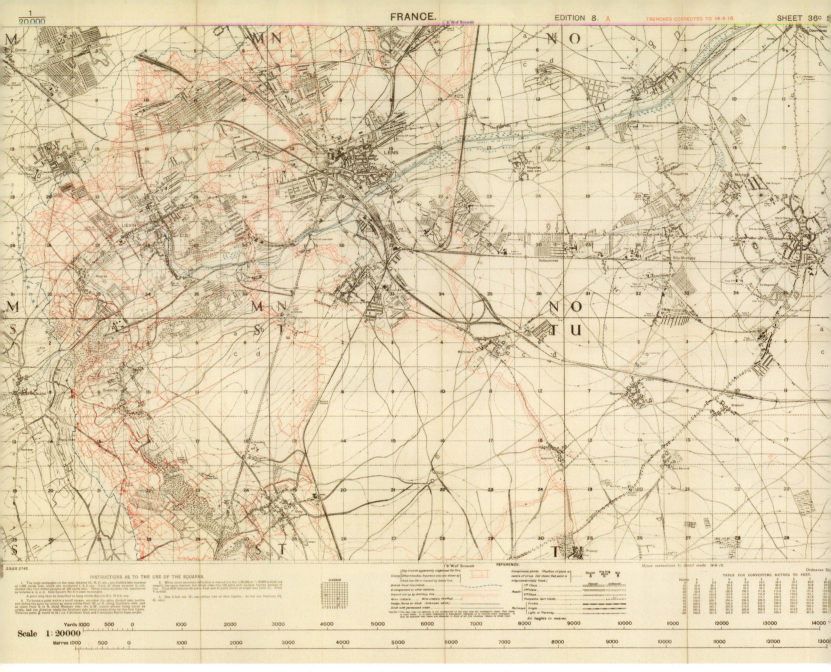

The 1:20,000-scale topographic map proved the most versatile for the Western Front. The scale could be enlarged to 1:10,000 or reduced to 1:40,000 without much loss of information. The legend on this 1:20,000-scale trench map shows some of the intelligence data that was tracked by British and Canadian units. Since the data was overprinted onto a base topographic map, it could be updated easily to meet the constantly changing situation along the Front.

Britain, Ordnance Survey, *Belgium and France, Trench Map, 1:20,000*, GSGS 2742, Ordnance Survey, 1915–18, sheet 36C SW, edition 8A, trenches corrected 14-09-16, NMC 114319

the entire base map each time. The information could be photographed, mapped, and distributed to units at the front within twenty-four to forty-eight hours.

The 1:20,000 scale was the most popular topographic map. It was considered the best scale for mapping the trenches and for showing special "target" and "position" information on German railways, dumps, machine-gun and mortar emplacements, wire entanglements, and observation posts. For this reason, the map was preferred by medium and heavy artillery, and for counter-battery work. The larger 1:10,000 map, on the other hand, was primarily intended as an infantry

trench map, and the smaller 1:40,000 map was used mostly in planning.

As far as the artillery was concerned, the one major improvement to the new map was the introduction of a common sheet system between France and Belgium. At the outbreak of the war, most European nations had not accepted Greenwich as the standard from which to measure 0° longitude, but used their own capital cities instead. Because of the lack of a standard, it was very difficult to match the topographic maps of one country with those of another. British mapmakers overcame this obstacle by simply extending the

Belgium system—which took Brussels as its base—into northeastern France.

The Allies used these sheet lines to develop a common grid overlay for all their topographic maps. The grid was a system of squares that could be used by the artillery to locate points on the map, a more accurate and easier referencing system than longitude and latitude. Since the same grid was used on all maps, anyone making reference to a target on one sheet would still be able to locate the reference point without having access to the same map, or even another map at the same scale.

In contrast, German maps of the Western Front used three different grids with origins based at Lille, Paris, and Rheims. Calculating the bearing of a gun when the target lay on a sheet mapped under one grid, the gun on the grid of another, and the nearest surveyed control point on yet a third grid must have caused enormous problems. But then German surveyors were not interested in making new topographic maps of the battlefields, relying instead on photographic enlargements of the early French hachured map. Even when the hachures were converted to contours, the Germans do not appear to have been concerned about accuracy. As the British surveyor Lieutenant Colonel Salmon reported after the war: "In the country between Le Cateau and LeQuesnoy maps were captured showing contours which were different from the French; and when we shortly afterwards took that country I was able to send out my own topographers to find out which were better—the French or German. They found the German contours entirely wrong." Apparently target information on German maps was not any different. Salmon could also provide examples where targets were as much as 1,100 metres out of position, and in one case an entire village was misplaced more than 100 metres.

Andrew McNaughton encountered similar inaccuracies in German mapping. "During the four weeks or so that we were on the Somme I don't suppose there was half an hour, day or night, in which a shell didn't land within a hundred yards of my headquarters. And occasionally there would be a hell of a concentration of fire and I was always frightened that they had got my guns spotted and were shooting at them." And so they had! His examination of captured German maps after the Somme found his guns marked, but in the

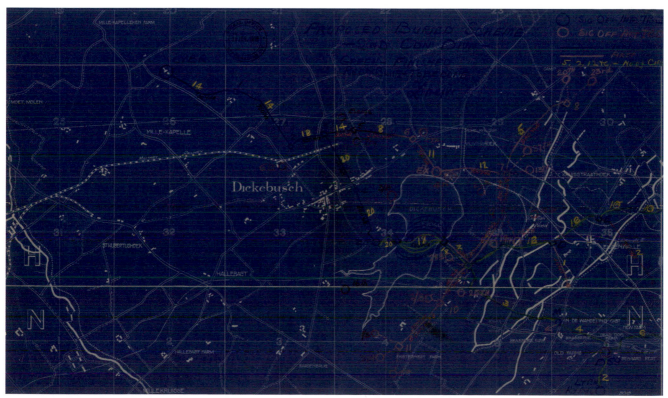

Communication with units at the front was vital to the success of any campaign. Maps allowed officers to plan the efficient distribution of roads, railway lines, and in this case, buried telephone cable.

Canada, Canadian Expeditionary Force, *Proposed buried cable scheme at Dickebusch, 2nd Canadian Division*, Signal Office, 2nd Canadian Division, 1916, NMC 115650

This kite balloon over Arras, France, helped mapmakers to collect vital target information on the German lines. A heavy metal cable tethered the balloon to a ground-based mobile winch that would haul the highly inflammable hydrogen-filled balloon and its observer back to the launch site once it attracted enemy fire.

Canada, Dept. of National Defence, *A kite balloon over Arras*, 1917, PA 003651

wrong place. "It was a case of locations put down on an intelligence map and never corrected," concluded McNaughton.

In the days leading up to the Canadian assault on Vimy, McNaughton and his staff worked feverishly with their new maps ciphering any intelligence that would help them nail down German targets. They paid particular attention to the new sciences of flash spotting and sound ranging, both of which required the placement of ground observers in the forward areas.

Flash spotters used surveying instruments to calculate the bearing of a gun flash. When two or more spotters fixed their instruments on the same flash, the point at which their bearings intersected would mark the location of the gun. When used under proper conditions, such methods of triangulation could locate a German gun to an accuracy of 4.5 metres. Sound rangers, on the other hand, used special instruments to pick up the sound of a gun discharge. Knowing the speed at which sound travels, the sound rangers could

estimate a gun position by measuring the amount of time it took for the sound of the discharge to pass over two or more fixed positions. Needless to say, both procedures were most effective when only one German gun was firing at a time. Under a heavy bombardment, it was often very difficult for an observer to know whether he was fixing his instruments on the same gun as his compatriots.

Although McNaughton tried to exploit all available methods of intelligence gathering, he found that almost a third of his counter-battery work came from information supplied by aerial photographs, specifically from photographs taken by pilots. In the earlier years of the war, photographs collected from balloon observation had been important, but as the war progressed and long-range artillery became more effective, balloons had to be placed too far from the front for obtaining accurate information. Since anti-aircraft fire was forcing observation aircraft to fly at greater heights, making pilot observation difficult, photographs taken from high-flying aircraft were the most important intelligence source.

Despite his best intentions, however, McNaughton's task was not always met with the enthusiasm it deserved. Just as Winterbotham had discovered before him, "[a] lot of the old traditional gunners, when they found these sorts of things were coming in and that people were trying to develop accuracy in shooting were appalled. Anybody who did it was looked on as somebody who ought to have his head read. This wasn't war at all," wrote McNaughton, "this was some sort of fandoodle."

The idea that you could pinpoint an enemy gun and knock it out of business was considered foul play by the old school of British gunners. As one officer once commented to Harold Heming, a McGill University graduate who was working with McNaughton in flash spotting, "you surveyors with your angles, co-ordinates and logarithms—you take all the fun out of war." Artillery was supposed to fire at the opposing infantry, not at each other.

With the help of the new maps and McNaughton's experiments in "scientific gunnery," the Canadians were able to fine-tune their methods of map shooting to the extent that creeping barrages in support of infantry offensives were possible. By making allowances for changes in the terrain, wind, weather, and barrel wear, their lumbering guns were transformed into sensitive weapons capable of shooting over the heads of the attacking troops. The curtain of exploding shrapnel dropped in front of the infantry was like a moving shield, unlike anything ever heard or seen before on the

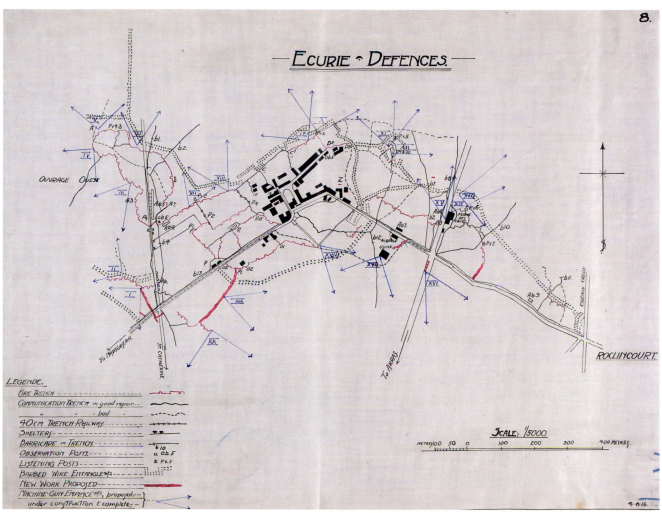

The fortified position at Écurie was among the many large-scale specialty maps produced by the Canadian Corps Topographic Section for the assault on Vimy. Such plans gave commanding officers a good overview of their defence situation.

Canada, Canadian Expeditionary Force, *Plan of Ecurie defences*, 3rd Army Troops Company, Canadian Engineers, 1916, NMC 112621

SUN PRINTING

Many of the base maps that Andrew McNaughton used at Vimy were prepared by Britain's 1st Field Survey Battalion, Royal Engineers, under the supervision of Capt. Herbert J.S. Gaine, a Dominion Land Surveyor from Vancouver. These maps were usually published by the Ordnance Survey on their massive printing presses at Southampton. The lithographic hand presses, which the Canadian Corps Topographic Section operated, could accommodate only the smaller print runs required for plans of redoubts, batteries, and other units of defence, as well as special operation maps and barrage sheets. Sometimes the Topographic Section produced the entire map themselves, but usually they relied on maps published by the Ordnance Survey and simply overprinted their own information.

Heavy lithographic stones and printing presses were awkward in the battlefield and lacked the mobility a war demanded. For this reason, the Canadians turned to blueprinting for their frontline units. First developed by the English astronomer Sir John Herschel in the 1840s and used commercially in North America during the Klondike

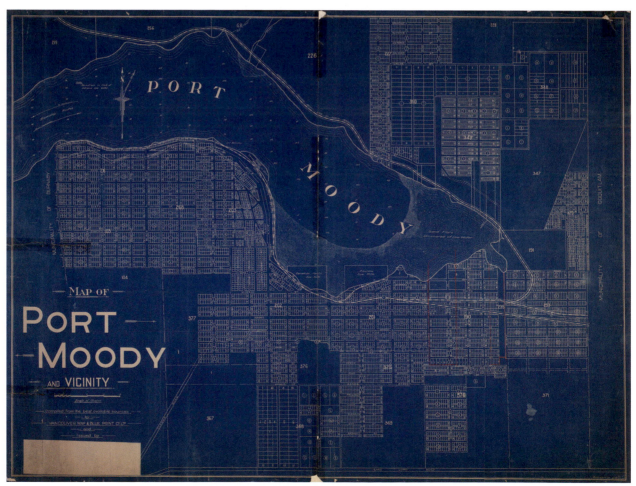

In the years immediately before the Great War, a number of map publishers in Canada's west coast communities used blueprinting technology. The Vancouver Map and Blue Print Company produced this cadastral map of Port Moody around 1908.

British Columbia, Dept. of Lands and Works, *Map of Port Moody and vicinity*, Vancouver, Vancouver Map and Blue Print Co., 1908–12, NMC 28924

SQUARE "O"

·DETAIL·OF·
·NO·MAN'S·LAND·
·IN·FRONT·OF·
·2ND· ·CANADIAN· ·DIVISION·

While engaged in the Ypres salient in 1916, the 2nd Canadian Division used sun printing to produce blueprint maps of the Front. The technology enabled officers to communicate information on no-man's-land without having to rely on special printing presses or tradesmen.

Canada, Canadian Expeditionary Force, *Detail of no-man's-land in front of 11th Canadian Inf Brigade*, 1916, NMC 21462

gold rush, the technology was favoured by a number of west coast map publishers by the outbreak of the war. Blueprinting called for a map to be drawn on translucent tracing paper, then placed in contact with paper pre-washed with a mixture of ferric ammonium citrate and potassium ferricyanide to make it light sensitive. Exposing the paper to the sun's ultraviolet light caused the ferric salts to turn an insoluble Prussian blue in areas not obscured by lines in the original drawing. The copy is at the same scale as the original map, but the image is a negative (white lines on a blue background). Once the exposed paper has been cleaned with water and dried, it remains reasonably robust providing it avoids exposure to excessive sunlight for long periods. On a sunny day the process took about twenty minutes, but in inclement weather, "sun printing" could take several days to make only a dozen copies of a single map.

Although the dark blue background made it difficult to add notes or make changes to the copy, sun printing did not require special machinery or a skilled printer and consequently was an efficient, cost-effective way for the Canadian Topographic Section to reproduce maps at the front. Although mechanical photocopying methods have largely replaced blueprints, the technology is still used in some contexts to reproduce large-format engineering and architectural drawings.

Aerial photo crews like this one with their *Vickers Viking IV* aircraft at Victoria Beach, Manitoba, in 1924, accelerated the mapping of Canada's northern frontier after the Great War.

Canada, Dept. of National Defence, *Crew of Vickers 'Viking' IV aircraft ...*, 1924, PA 053239

Western Front. To many of the infantry, the ear-shattering barrage was so intense that it seemed as if they were advancing behind a solid wall. "I felt that if I lifted a finger I should touch a solid ceiling of sound," recalled one Canadian private. The tactic proved effective; it paralyzed the Germans, forcing them to remain hidden in their bunkers just when the attacking Canadian troops were in greatest danger.

The success at Vimy led to the formation of the Canadian Corps Survey Section about a year later under Capt. W.R. Flewin, and ultimately laid the foundation for the introduction of Canadian artillery survey units a few years after the war. In addition, it encouraged the Allies to place an even greater emphasis on their mapping program. With the development of aerial photography, the project proceeded at a pace that had been unimaginable in the pre-war years. By the end of the war, mapping units along the Western Front had grown into a work force of some 1,200 personnel and had produced more than thirty-two million map sheets, the largest production the world had ever witnessed for a single region.

Interestingly, despite these enormous production figures, many combatants had no idea of the effort that the Allies put into mapmaking. Even field officers were frequently taken aback by the number of maps required for a single campaign. The receiving officer for one consignment that had been put together for the Front wanted to see the maps simply handed over so that he could go about his regular business. "I'll put them in my pocket and get away with them right away," was his comment. He was quite surprised on hearing that the shipment consisted of several thousand sheets, weighed more than a ton, and would require a railway box car!

Even though one Royal Artillery officer, Major General Franks, later described the advances made in map shooting as "one of the wonders of the war," only a small percentage of the trench maps used in the

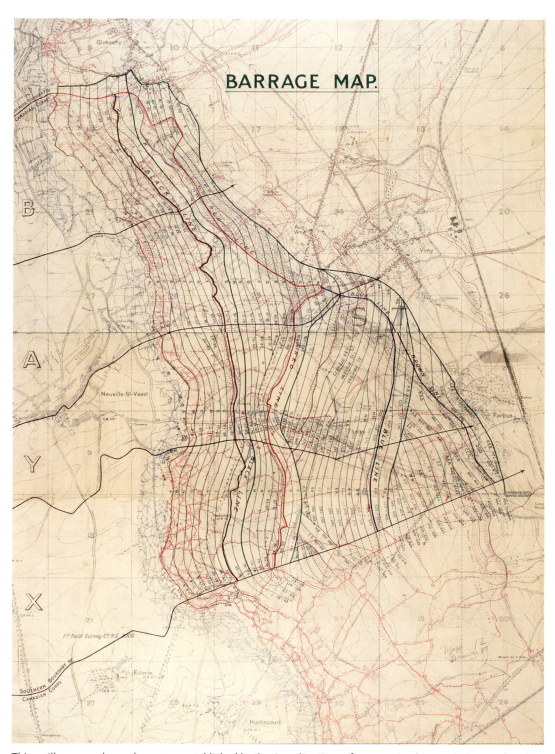

BARRAGE MAP.

This artillery map shows the pattern established by the Canadian Corps for a creeping barrage in its attack on Vimy in April 1917. The timings printed next to the heavy coloured lines indicated the points at which artillery fire must lift off to allow the infantry to move forward. The map also indicates divisional boundaries and the destination points of successive waves of attacking infantry (black, red, blue, and brown).

Canada, Canadian Expeditionary Force, *Barrage map showing boundaries and objectives for assault on Vimy Ridge*, 1st Field Survey Company, 1917, NMC 111113

Great War have survived. This outcome is most unfortunate because, unlike other documents of the First World War, maps alone provide an important visual record of the obstacles and hardships faced by front-line soldiers. More significant for Canadians, maps of the Western Front stand as testaments to our own unique scientific and technical contributions to the war effort.

This computer-generated and hand-painted map of Canada provides a bird's-eye perspective of the country as it would appear from an earth-observation satellite stationed above the equator. Canada plays a leading role in developing remote sensing systems; RADARSAT, our latest contribution, was launched in November 1995. It was the first satellite to be equipped with a powerful Synthetic Aperture Radar (SAR) instrument capable of acquiring images of the earth, day or night, and through cloud cover, smoke, and haze.

Canada. National Film Board, Ann Kask and Lorne Kask (artists), *NFB Canada map*, 1984, NMC 131999

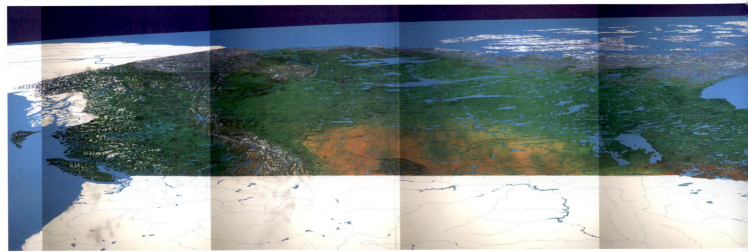

Conclusion

Terra Nostra is a celebration of the mapping of Canada—for Library and Archives Canada, it is also a celebration and a sharing of the extraordinary historical resource that is contained in our cartographic holdings. From the inception of our institution in 1872 until today's Library and Archives Canada, part of our mandate has been to acquire and preserve the cartographic heritage of Canada. The collection spans a broad range of mapping initiatives, from the portolan charts of the early explorers, to recent aerial images of east coast ice floes.

Archivist Hensley R. Holmden's published inventory of the holdings from 1912 counted some 4,100 maps in the collection. Today, almost a century later, our maps, plans, and charts number closer to 1.7 million items. Taken in its entirety, the collection constitutes the world's single largest description of our country's landscape. Thanks to a very active acquisitions program, our cartographic legacy will continue into the future. The legal deposit initiative guarantees that Library and Archives Canada will acquire modern, up-to-date maps, charts, and atlases from private publishers and government agencies systematically, as they come off the press. Many of these newly acquired items will become tomorrow's cartographic treasures. Our state-of-the-art preservation facility will ensure that all these records will be available for the use and enjoyment of generations to come.

To some extent, this book also celebrates Canadian ingenuity in the surveying and mapping professions. Faced with an almost insurmountable landscape, Canadian mapmakers were quick to exploit new technologies to speed up the mapmaking process and distribute their wares in ever-increasing numbers, at ever-decreasing prices, to an ever-growing and geographically literate population. The immigration atlases were examples of one of the earliest federal government uses of the cartographic medium in support of national aspirations. County map providers, bird's-eye artists, and road map producers contributed to a fledgling private industry, whose workforce today employs an estimated 30,000 Canadians and contributes about $4 billion a year to the national economy in business revenue.

Ever since the first box cameras were hauled up the slopes of the Canadian Rockies more than a century ago, Canada has continued to lead in the automated mapping field—especially in the development of Geographical Information Systems (GIS). These computer-based systems store maps in digital format, which

allows them to be read, measured, analyzed, and manipulated through sophisticated software. The idea of using computers in mapmaking was first conceptualized by Roger Tomlinson in the early 1960s. Under his leadership, and with the support of the Department of Mines and Technical Surveys (the forerunner of today's Natural Resources Canada), Tomlinson laid the groundwork for later developments in GIS around the world. Through GIS technology, it is now possible to map almost anything and to carry these maps in the palm of a hand and on the dashboard of a car. Its use allows greater understanding of human interactions and their impact on natural environments. GIS plays an important role in today's decision making at all levels of government and in all sectors of business.

More recently, Canadian ingenuity in mapmaking has manifested itself through the RADARSAT program. Built at a cost of $620 million, RADARSAT is a sophisticated earth-observation satellite, developed by the Canada Centre for Remote Sensing, the Canadian Space Agency, and a consortium of one hundred other government and private sector organizations. Using Synthetic Aperture Radar (SAR) instead of natural light, the satellite is capable of seeing through the darkness and cloud cover that enshroud our Arctic regions for much of the year. Although the type of SAR system used was difficult to implement technically, it has proved extremely versatile in helping to distinguish between different kinds of sea ice, and in measuring the density and depth of snow. With such capabilities,

RADARSAT has proved itself very effective as a tool in monitoring ice floes throughout Canada's northern shipping routes and offshore drilling sites. RADARSAT imagery is also used to measure the moisture content in soil and vegetation, to survey the health of crops and forests, to detect oil spills and forest fires, and to forecast floods. Circling the earth by the poles fourteen times a day, at an altitude of 832 kilometres, RADARSAT travels in a sun-synchronous orbit—the plane of the orbit does not move relative to the direction of the sun. In other words, the satellite is set to cross the equator at the same place and at the same local time every twenty-four days. The second generation of RADARSAT will soon be operational. It will have improved spatial resolutions ranging from 3 to 100 metres, and will be capable of covering a 10- to 500-kilometre path across the earth's surface.

Canada's maps have been many things to many people: for early merchant-adventurers, they promised fame and fortune; for army commanders, they strengthened their hold on the land; for politicians, they legitimized their dominion over the country; for businessmen, they proclaimed their sense of community pride; and for ordinary citizens, they were testaments to the type of communities they wanted for their children. As repositories of intellectual thought, Canada's maps provide evidence of our nation's technical accomplishments and offer insight into the growth of our geographical knowledge. They are windows on our view of the world. ⌘

Selected Readings

SECTION I

Envisioning Canada

Maps have played a fundamental role in helping us to define Canada. For a general discussion of some key historical maps (and other archival records) found in the Library and Archives Canada collection, see the four works published in the series *Records in Our History*: André Vachon, *Dreams of Empire: Canada before 1700* (Ottawa: Public Archives Canada, 1982); André Vachon, *Taking Root: Canada from 1700 to 1760* (Ottawa: Public Archives Canada, 1985); Bruce G. Wilson, *Colonial Identities: Canada from 1760 to 1815* (Ottawa: National Archives of Canada, 1988); and George Bolotenko, *A Future Defined: Canada from 1849 to 1873* (Ottawa: National Archives of Canada, 1992). A highly colourful general work focusing on the collection can also be found in *Treasures of the National Archives of Canada* (Toronto: University of Toronto Press, in cooperation with the National Archives of Canada, 1992).

A good description of various map printing technologies through the ages can be found in David Woodward, *Five Centuries of Map Printing* (Chicago: University of Chicago Press, 1975).

1 Passage to the Orient

Library and Archives Canada has a number of original records in a variety of media—both published and unpublished—that were created in connection with the search for a Northwest Passage. These works are scattered throughout the institution's holdings and are accessible through the creator's name.

For a general history of the search for a Northwest Passage through Arctic waters, see James P. Delgado, *Across the Top of the World: The Quest for the Northwest Passage* (Vancouver: Douglas and McIntyre, 1999); Seymour I. Schwartz, "Depicting a Desire," in *The Mismapping of America* (Rochester: University of Rochester Press, 2003), pp. 77–126; and J.L. Allen, "The Indrawing Sea: Imagination and Experience in the Search for the Northwest Passage, 1497–1632," in *American Beginnings*, edited by E.W. Baker (Lincoln: University of Nebraska Press, 1996), pp. 7–35. For a detailed history of the British explorations that followed the Mackenzie voyages, see Pierre Berton, *The Arctic Grail: The Quest for the North West Passage and the North Pole, 1818–1909* (Toronto: McClelland and Stewart, 1988). For an academic treatment of the French search for a passage, see Lucie Lagarde, "Le Passage du Nord-Ouest et la Mer de l'Ouest dans la cartographie française du 18ᵉ siècle,

contribution à l'étude de l'œuvre des Deslisle et Buache," *Imago Mundi*, no. 41 (1989), pp. 19–43. For a lavishly illustrated history of Samuel de Champlain as cartographer, explorer, and governor of New France, see Raymonde Litalien and Denis Vaugeois, eds., *Champlain: The Birth of French America* (Montréal: McGill-Queen's University Press, 2004). For Champlain's published works, see Samuel de Champlain, *The Works of Samuel de Champlain*, 6 vols. (Toronto: Champlain Society, 1922–36).

2 General Murray Maps the St. Lawrence

Only seven copies of the Murray map were made, six of which were distributed as follows: one went to King George III; one to the British prime minister, William Pitt; one to the newly formed Board of Ordnance; one to the governor general, Jeffrey Amherst; one to Gage; and one seems to have been kept in Québec by General Murray himself. Of the seven copies produced, only five have been traced, two of which are at Library and Archives Canada, two at the British Museum, and one at the William Clements Library, University of Michigan. For a general description of the five known copies, see N.N. Shipton, "General James Murray's Map of the St. Lawrence," *The Cartographer* 4, no. 2 (1967), pp. 93–102, and J.G. Shields, "The Murray Map Cartographically Considered" (M.A. thesis, Queen's University, 1980). For a detailed social history of British military engineers at the time of the Seven Years War, see Douglas William Marshall, "The British Military Engineers 1741–1783: A Study of Organization, Social Origin and Cartography" (Ph.D. diss., University of Michigan, 1976).

3 "Free Farms for the Million"

Most of the English-language editions of the immigration atlases and some French-language editions can be found in the Library and Archives Canada collection. Regrettably, the editions that were published in other European languages are not well represented. The existence of some editions is known only through the listings of documents that government departments published in their annual reports. The English-language editions of the atlas were published, at various times, under the titles *Canada West*, *The Last Best West*, *A Descriptive Atlas of Western Canada*, and *Canada: Descriptive Atlas*. The French-language editions were titled *Géographie du Canada et atlas de l'ouest canadien*, *Canada: atlas descriptif*, and *Atlas de l'ouest canadien*. Before 1917, the atlases were published by the Immigration Branch of the Department of the Interior; after 1917, publication was taken over by the Department of Immigration and Colonization.

Canada's research community has never considered immigration advertising a serious field of study and no detailed histories have been written. For a more complete discussion of the settlement process, see Pierre Berton, *The Promised Land: Settling the West 1896–1914* (Toronto: Anchor Canada, 2002); Jean Bruce, *The Last Best West* (Toronto: Fitzhenry and Whiteside, in association with the Multiculturalism Programme, Department of the Secretary of State, 1976); and Ronald Ress, *New and Naked Land: Making the Prairies Home* (Saskatoon: Western Producer Prairie Books, 1988). For a history of the western homestead surveys, see James G. MacGregor's *Vision of an Ordered Land: The Story of the Dominion Land Survey* (Saskatoon: Western Producer Prairie Books, 1981).

SECTION II
Perfecting Our Cities

Canada's urban cartography is highlighted in two specialized works: Alan F. J. Artibise and Edward H. Dahl, *Winnipeg in Maps, 1816–1972* (Ottawa: Public Archives of Canada, 1975), and Thomas L. Nagy, *Ottawa in Maps: A Brief Cartographical History of Ottawa, 1825–1973* (Ottawa: Public Archives Canada, 1974). Both these works highlight archival records found in the Library and Archives Canada collection and provide readers with a good overview of the different types of urban maps produced over the years by Canadian agencies. Canada's urban photography is celebrated in Lilly Koltun, *City Blocks, City Spaces: Historical Photographs of Canada's Urban Growth, c. 1850–1900* (Ottawa: Public Archives Canada, 1980). Again, this work focuses on archival records found in the Library and Archives Canada collection.

Urban mapping has been the subject of two British exhibitions, both of which published catalogues. See James Elliot, *The City in Maps: Urban Mapping to 1900* (London: The British Library, 1988), and Nick Millea, *Street Mapping: An A to Z of Urban Cartography: An Exhibition in the Bodleian Library, February—April 2003* (Oxford: Bodleian Library, University of Oxford, 2003). The urban cartography of Britain is also presented in Ashley Baynton-Williams, *Town and City Maps of the British Isles, 1800–1855* (Salford: Dolphin Publications, 1993). For a more academic treatment of urban cartography, see David Buisseret, ed., *Envisioning the City: Six Studies in Urban Cartography* (Chicago: University of Chicago Press, 1998).

4 Modelling Québec

The Duberger and By model of Québec can be viewed by the public at Artillery National Historic Park, Québec. Specialized studies of the model have been published by André Charbonneau, *The Model of Quebec*, The Fortifications of Quebec Series, Booklet No. 1 (Ottawa: Parks Canada, 1981),

and by Bernard Pothier, *The Quebec Model* (Ottawa: National Museums of Canada, 1978).

Louis XIV's model collection is also on public display. Various parts of it can be seen at the Hôtel des Invalides in Paris. For a history of early model building by the French, see Jeffrey S. Murray, "Princely Toys: Louis XIV's Collection of Meticulously Crafted Models of Cities Offers Modern Historians a Unique Perspective on Long-Vanished Urban Landscapes," *MHQ: The Quarterly Journal of Military History* 14, no. 2 (2002), pp. 80–85; Catherine Brisac, *Le musée des plans-reliefs: Hôtel national des invalides* (Paris: Pygmalion / Gérard Watelet, 1981); J. Grignard, *Musée des plans reliefs* (Paris: Hôtel des Invalides, 1969); A. de Roux, N. Faucherre, et al., *Les plans en relief des places du roy* (Paris: A. Biro, 1989); and Honoré Bernard, Philippe Bragard, Nicolas Faucherre, et al., *Plans en relief: villes fortes des anciens Pay-Bas français au XVIIIe siècle* (Lille: Musée des Beaux-Arts de Lille, 1989).

5 A Bird's-Eye Perspective

The largest collection of bird's-eye views of Canadian cities is found in the holdings of Library and Archives Canada. Over the years the views have been acquired and catalogued by several different areas of the institution—some are classified as documentary art, for example, and others as cartographic. Unfortunately, Library and Archives Canada has no single source that itemizes its holdings of bird's-eye views, and without knowing either the title or creator of a view, researching the collection can be very difficult.

The most complete secondary source on bird's-eye views is John W. Reps, *Views and Viewmakers of Urban America: Lithographs of Towns and Cities in the United States and Canada, Notes on the Artists and Publishers, and a Union Catalog of Their Work, 1825–1925* (Columbia: University of Missouri Press, 1985). Although Reps is primarily interested in the American industry, there is some minor discussion of Canadian views throughout. A very beautiful presentation of the bird's-eye views of over a hundred American and nine Canadian cities can be found in John W. Reps, *Bird's Eye Views: Historic Lithographs of North American Cities* (New York: Princeton Architectural Press, 1998). Other than the discussion presented in this volume, no historical treatments of Canadian bird's-eye views has been undertaken.

6 Mapping for Fire

Goad's fire insurance plans (and those of his successors) are scattered among archival collections and libraries across the country. An inventory of the plans in 122 major Canadian repositories was published by Lorraine Dubreuil and Cheryl A. Woods, *Catalogue of Canadian Fire Insurance Plans, 1875–1975* (Ottawa: Association of Canadian Map Libraries and Archives, 2002).

The single largest collection of Canadian insurance plans can be found in the holdings of Library and Archives Canada;

they are partially inventoried in Robert J. Hayward, *Fire Insurance Plans in the National Map Collection* (Ottawa: Public Archives Canada, 1977). An equally impressive collection of Canadian insurance plans can be found in the British Library. The plans were originally sent to the British Museum between 1895 and 1923 as part of the requirements for registration under Canada's *Copyright Act* and at some point were transferred to the Library. A listing of these holdings can be found in Patrick B. O'Neill, *Checklist of Canadian Copyright Deposits in the British Museum, 1895–1923*, vol. 2, *Insurance Plans* (Halifax: Dalhousie University, School of Library Service, 1985). The Canadian plans were colour microfilmed (105 mm format) in the 1990s by Library and Archives Canada and are now available for consultation in its reference room.

Two inventories of the insurance plans available in provincial collections have been compiled by France Woodward, "Fire Insurance Plans and British Columbia Urban History: A Union List," *BC Studies*, no. 42 (Summer 1979), pp. 13–49, and by Marcel Fortin, Lorraine Dubreuil, and Cheryl A. Woods, *Canadian Fire Insurance Plans in Ontario Collections, 1876–1973* (Ottawa: Association of Canadian Map Libraries and Archives, 1995).

SECTION III
Finding Our Way

For a highly readable account of some of the major maps that have helped us to navigate our way across Canada, see Alan Morantz, *Where Is Here?: Canada's Maps and the Stories They Tell* (Toronto: Penguin Canada, 2002). Various maps of exploration and route building are also discussed in Don W. Thomson, *Men and Meridians: The History of Surveying and Mapping in Canada*, 3 vols. (Ottawa: R. Duhamel / Queen's Printer, 1966).

7 Charting Eastern Waters

The importance of *The Atlantic Neptune* to the maritime history of North America is undisputed, yet there has been no attempt to systematically inventory all the various editions of Joseph F.W. Des Barres's charts. Without such an inventory, the full extent of his prolific work can only be surmised. The single largest archival collection of Des Barres's records is found at Library and Archives Canada, where some seven hundred different editions and states of his original charts of Canadian waters are preserved along with more than forty colourful coastal views and profiles that were published separately from the charts. Some of the charts are described in Kenneth A. Kershaw, *Early Printed Maps of Canada*, vol. 1 (Ancaster, Ont.: Kershaw Publishing, 1993). Several of the views can be found in Richard L. Raymond, *J.F.W. DesBarres: Views and Profiles* (Halifax: Dalhousie Art Gallery, 1982). Library and Archives Canada also retains the Des Barres papers, consisting of 2.5 metres of original correspondence,

memoranda, accounts, and land records. The records span his entire career as a mapmaker, public servant, and land owner in Nova Scotia.

Library and Archives Canada has the largest collection of rare original copper plates from the last edition of *The Atlantic Neptune*. The thirty plates were acquired in 1946 from a collection of sixty-four still in the possession of the British Admiralty (at one time the number of copper plates in the Admiralty's stores numbered 287, but the majority of these were sold in the nineteenth century as scrap). Thirty-four of the remaining plates—those covering American waters—were distributed by the Admiralty to twelve separate institutions in the United States.

Des Barres's career, beyond that of hydrographer for the British Admiralty, is discussed in Geraint N.D. Evans, *Uncommon Obdurate: The Several Public Careers of J.F.W. DesBarres* (Toronto: University of Toronto Press, 1969), and in John Clarence Webster, *The Life of Joseph Frederick Wallet DesBarres* (Shediac, N.B.: privately printed, 1933). Historical accounts of *The Atlantic Neptune* are provided by Christopher Terrell in "The Magnificent Atlantic Neptune," *The Geographical Magazine* 53, no. 15 (1981), pp. 956–61, and "A Sequel to *The Atlantic Neptune* of J.F.W. Desbarres: The Story of the Copperplates," *The Map Collector*, no. 72 (1995), pp. 2–9.

Readers wishing to put Des Barres's work in the broader context of British hydrographic mapping should consult G.S. Ritchie, *The Admiralty Chart: British Naval Hydrography in the Nineteenth Century* (Edinburgh: Pentland Press, 1995). More general treatments of worldwide chartmaking through the ages can be found in Peter Whitfield, *The Charting of the Oceans: Ten Centuries of Maritime Maps* (Rohnert Park, Calif.: Pomegranate Artbooks, 1996), and in Derek Hayes, *Historical Atlas of the North Pacific Ocean: Maps of Discovery and Scientific Exploration 1500–2000* (Vancouver: Douglas and McIntyre, 2001).

For an overview of Canadian hydrographic mapping, see Stanley Fillmore and Robert W. Sandilands, *The Chartmakers: The History of Nautical Surveying in Canada* (Toronto: NC Press, 1983), and William Glover, ed., *Charting Northern Waters: Essays for the Centenary of the Canadian Hydrographic Service* (Montréal: McGill-Queen's University Press, 2004).

8 Going for Klondike Gold

Library and Archives Canada has an extensive collection of original maps, documentary art, and original photography relating to the Klondike gold rush. The maps are accessible by searching under the creator's name. The unpublished Alphonso Waterer manuscript entitled "The Great Loneland" is found in MG 30 C59.

With regard to secondary sources, Pierre Berton, *Klondike: The Last Great Gold Rush, 1896–1899* (Toronto: McClelland and Stewart, 1972) is still the most complete account of the great stampede to the Yukon. Although he does not discuss any of the maps or other literature that promoted the gold rush, Berton's history of the gold rush makes for first-rate reading. More recently, see David Neufeld and Frank Norris, *Chilkoot Trail: Heritage Route to the Klondike* (Whitehorse: Lost Moose, 1996) for a colourful and well-illustrated history of the most popular route taken by the stampeders in 1897–98. Tappan Adney, *The Klondike Stampede* (Vancouver: University of British Columbia Press, 1994) is a wonderful first-hand account of the great stampede by one of its participants. Originally published at the turn of the century, it has long been considered a classic of Yukon gold rush literature. Finally, an account of one of the surveyors who helped keep a semblance of sanity in the gold fields can be found in Don W. Thomson, "The Yukon Gold Rush and William Ogilvie, DLS," in *Men and Meridians: The History of Surveying and Mapping in Canada*, vol. 2, *1867–1917* (Ottawa: R. Duhamel / Queen's Printer, 1966), pp. 147–61. With the exception of this chapter, no treatments of the gold rush mapping industry have been undertaken by historians.

9 Maps for Your Motoring Pleasure

Although the road maps handed out by local gas stations and provincial governments are probably best known to Canadians, in reality this type of mapping was created by all levels of government and by a variety of commercial publishers other than the major oil companies. Although far from complete, Library and Archives Canada has a number of the maps produced by government and non-government agencies. These works are distributed throughout the institution's holdings and are accessible by searching under the name of the creator.

Other than this chapter, there are no known historical treatments of Canadian road maps. Some excellent studies on the American aspects of the industry have been published, however. Most notable among these are Walter W. Ristow, "A Half Century of Oil-Company Road Maps," *Surveying and Mapping* 24, no. 1 (1964), pp. 617–37, and Douglas E. Yorke, J. Margolies, and E. Baker, *Hitting the Road: The Art of the American Road Map* (San Francisco: Chronicle Books, 1996). British road maps are considered in T.R. Nicholson, *Wheels on the Road: Maps of Britain for the Cyclist and Motorist, 1870–1940* (Norwich: Geo Books, 1983). French road maps are discussed in B. Paternault, "Les cartes touristiques Michelin," *Bulletin du Comité français de cartographie* 115, no. 1 (1988), pp. 33–36. As well, the recently launched quarterly *Route nostalgie* regularly examines all aspects of France's automobile industry, including its map production.

For a good introduction to the creation of Canada's highway network, see Don W. Thomson, "Evolution of Canadian Highway Survey," in *Men and Meridians: The History of Surveying and Mapping in Canada*, vol. 3, *1917–1947* (Ottawa: R. Duhamel / Queen's Printer, 1966), pp. 133–54.

SECTION IV
Scaling the Landscape

There are a number of excellent secondary sources on the early cadastral and topographical mapping of Canada, in particular, W. D. Stretton, "From Compass to Satellite: A Century of Canadian Surveying and Surveyors 1882–1982," *The Canadian Surveyor* 36, no. 4 (1982), pp. 1–356; Norman L. Nicholson and Louis M. Sebert, *The Maps of Canada: A Guide to Official Canadian Maps, Charts, Atlases and Gazetteers* (Folkestone, England: Wm. Dawson and Sons, 1981); Don W. Thomson, *Men and Meridians: The History of Surveying and Mapping in Canada*, 3 vols. (Ottawa: R. Duhamel / Queen's Printer, 1966); and John L. Ladell, *They Left Their Mark: Surveyors and Their Role in the Settlement of Ontario* (Toronto: Dundurn Press, 1993). More recent mapping—the period after the Second World War in which Canada distinguished itself in remote sensing and the development of geographical information systems—can be found in Gerald McGrath and Louis M. Sebert, *Mapping a Northern Land: The Survey of Canada, 1947–1994* (Montréal: McGill-Queen's University Press, 1999).

10 Hawking County Maps

The county maps and atlases in the Library and Archives Canada collection are partly described by Heather Maddick in *County Maps: Land Ownership Maps of Canada in the 19th Century* (Ottawa: Public Archives of Canada, 1976), and by Betty May in *County Atlases of Canada: A Descriptive Catalogue* (Ottawa: Public Archives of Canada, 1970).

Canada's most prolific county mapmaker, Ambrose F. Church, is the subject of two detailed studies by C.B. Fergusson, "Ambrose F. Church, Map-Maker," *The Dalhousie Review* 49, no. 4 (1969), pp. 505–16, and "Ambrose F. Church and His Maps," *Journal of Education* (Province of Nova Scotia) 19, no. 4 (1970), pp. 19–27.

The central source on nineteenth-century door-to-door canvassing is *How 'Tis Done: A Thorough Ventilation of the Numerous Schemes Conducted by Wandering Canvassers Together with the Various Advertising Dodges for the Swindling of the Public* (Syracuse: W.I. Pattison, 1890). Although the title was published anonymously, most scholars agree that it is the work of Bates Harrington. Essential secondary sources on the industry are Jeffrey S. Murray, "A Gift for the Gab: Hawking Books in the Hinterland," *Biblio* 3, no. 7 (1998), pp. 14–16, and "Door-to-Door Bookstores," *Legion Magazine* 78, no. 4 (2003), pp. 32–34, as well as David Bosse, "A Canvasser's Tale," *The Map Collector*, no. 57 (1991), pp. 22–26. Readers interested in the industry should also look at Keith Arbour, *Canvassing Books, Sample Books, and Subscription Publishers' Ephemera 1833–1951 in the Collection of Michael Zinman* (Ardsley, N.Y.: The Haydn Foundation for the Cultural Arts, 1996).

11 Deville's Topographic Camera

Land surveying is a historical process in which each surveyor builds on the work of predecessors. It is to be expected, then, that the records relating to the development of photo-topography in Canada are wide-ranging and cover a variety of media. Library and Archives Canada, for example, holds all the original glass-plate negatives from the mountain surveys, all the maps that were published as a result of these surveys, and a good many of the correspondence files that the Department of the Interior and its successor agency, the Department of Mines and Resources, created in their administration of the mountain surveys. The majority of the surveyors' field notebooks—those used to record measurements and, next to the photographs, would be a principal source of information when compiling a map—were transferred by the Department of Energy, Mines and Resources and the Public Archives of Canada to their respective provincial agencies in the 1970s. The field notebooks pertaining to the survey of federal crown lands—Indian reserves, national parks, and so on—and international and inter-provincial boundaries are still retained by the Legal Surveys Division of Natural Resources Canada.

Despite the excellent archival record that has survived, there are few secondary sources on the history of photo-topography. The one exception is D.W. Thomson, "Deville and the Survey Camera," in *Men and Meridians: The History of Surveying in Canada*, vol. 2, *1867–1917* (Ottawa: R. Duhamel/Queen's Printer, 1966), pp. 131–46.

For a detailed description of the western homestead surveys undertaken by the federal government, see John Y. Tyman, *By Section, Township and Range: Studies in Prairie Settlement* (Brandon: Assiniboine Historical Society, 1972). Some of the maps that were produced as a result of the federal surveys are listed in G. Poulin and F. Cadieux, *Index to Township Plans of the Canadian West* (Ottawa: Public Archives of Canada, 1974). This index is now largely out of date, however, because of the active acquisitions program at Library and Archives Canada. A second western map series published by the federal government is described by Lorraine Dubreuil, *Sectional Maps of Western Canada, 1871–1955: An Early Canadian Topographical Map Series* (Ottawa: Association of Canadian Map Libraries and Archives, 1989). For a discussion of the development of aerial photography in Canada, see Don W. Thomson, *Skyview Canada: A Story of Aerial Photography in Canada* (Ottawa: Energy, Mines and Resources Canada, 1975).

The 1921 expedition to Everest, and many that followed, ended in failure, the summit having been first conquered by Edmund Hillary, a New Zealander, and Tenzing Norkay, a Nepalese Sherpa guide, on May 29, 1953. For a description of the 1921 expedition and a copy of the resulting map, see E.O. Wheeler, "The Mount Everest Expedition, 1921," *Canadian Alpine Journal* 13 (1922), pp. 1–25.

Édouard Gaston Deville's contribution to mapmaking has never been properly investigated. A small hint on the extent of his contribution can be surmised from "Edward [*sic*] Gaston Daniel Deville," *Proceedings and Transactions of the Royal Society of Canada* 19, 3rd ser. (1925), pp. viii–xi, and G.S. Andrews, "Edouard Gaston Daniel Deville, 'Father' of Canadian Photogrammetric Mapping," *The Canadian Surveyor* 30, no. 1 (1976), pp. 36–40. Deville's use of cameras in surveying is described more fully in his pamphlet *The Map of the Rocky Mountains Park of Canada and Surrounding Country and the Canadian Topographical Surveys Executed by Means of Photography* (Ottawa: S.E. Dawson, 1893), and in his manual *Photographic Surveying including the Elements of Descriptive Geometry and Perspective* (Ottawa: Government Printing Bureau, 1895).

12 Weapons of the Great War

Library and Archives Canada has a rich collection of trench maps from the First World War. Although the majority of these were created by British and Canadian units, maps by American, French, and German cartographers are also represented. Some 3,000 British-Canadian trench maps from thirty-two separate archival fonds have been intellectually collated and itemized in an unpublished finding aid. This guide is available to researchers in the main reference room of Library and Archives Canada. Canadian units often attached maps to their monthly diaries. These diaries and their map appendices are not listed in the guide to trench maps but are available to researchers on the Library and Archives Canada website.

The definitive history on British military cartography on the Western Front is Peter Chasseaud's monumental *Artillery's Astrologers: A History of British Survey and Mapping on the Western Front 1914–1918* (Lewes: Mapbooks, 1999). For a more condensed academic treatment of the maps produced by British and Canadian units, see Jeffrey S. Murray, "British-Canadian Military Cartography on the Western Front, 1914–1918," *Archivaria* 26 (Summer 1988), pp. 52–65, and Peter Chasseaud's *Trench Maps: A Collector's Guide* (Lewes: Mapbooks, 1986), and "British Artillery and Trench Maps on the Western Front 1914–18," *The Map Collector*, no. 51 (1990), pp. 24–32.

The Flemish cartographer Theodor de Bry mapped the western hemisphere in 1596 and surrounded it with four of Europe's notable explorers: Christopher Columbus, Ferdinand Magellan, Amerigo Vespucci, and Francisco Pizarro. The open Arctic waters across the top of Canada hinted at a Northwest Passage.

Theodor de Bry, *America sive novvs orbis respectv Evropaeorvm inferior globi terrestris pars*, Francofurti ad Moenum formis Theod. de Bry, 1596, NMC 27654

Index

References to illustrations are printed in boldface type.

DESIGNED AND TYPESET BY JOSÉE LALANCETTE
IN MINION 11.3 AND BERNHARD MODERN
THIS BOOK WAS PRINTED IN MARCH 2006
ON GARDA SILK PAPER 200M
BY FRIESENS
ALTONA MANITOBA
FOR GILLES HERMAN AND DENIS VAUGEOIS
LES ÉDITIONS DU SEPTENTRION